AF698102

Biodefence Mechanisms Against Environmental Stress

Biodefence Mechanisms Against Environmental Stress

Toshihiko Ozawa *National Institute of Radiological Sciences, Japan*

Kouichi Tatsumi *National Institute of Radiological Sciences, Japan*

Tada-aki Hori *National Institute of Radiological Sciences, Japan*

Kodansha

Springer

QH
465
B56
1998

Toshihiko Ozawa
Director
Department of Bioregulation Research
National Institute of Radiological Sciences
Chiba 263, Japan

Kouichi Tatsumi
Director
Division of Biology and Oncology
National Institute of Radiological Sciences
Chiba 263, Japan

Tada-aki Hori
Director
Genome Research Group
National Institute of Radiological Sciences
Chiba 263, Japan

Copublished by
KODANSHA LTD.
12-21 Otowa 2-chome, Bunkyo-ku, Tokyo 112, Japan
and
SPRINGER-VERLAG GmbH & Co. KG
Tiergartenstr. 17, D-69121 Heidelberg, Germany

exclusive sales rights in Japan:
Kodansha Ltd., Tokyo
worldwide except Japan:
Springer-Verlag, Heidelberg

ISBN 3-540-63899-7 Springer-Verlag Berlin Heidelberg New York
ISBN 4-06-208948-3 Kodansha

Copyright © National Institute of Radiological Sciences (NIRS), 1998
All rights reserved. No part of this book may be reproduced in any form, by photostat, microfilm, retrieval system, or any other means, without the written permission of NIRS (except in the case of brief quotation for criticism or review).

Printed in Japan

List of Contributors

Numbers in parentheses refer to the pages on which contributors' paper begins.

Ajimura, M. (115)
Genome Research Group, NIRS, Chiba 263, Japan

Berg, R.J.W. (181)
Dept. of Dermatology, University of Utrecht, Utrecht, The Netherlands

Boiteux, S. (107)
Commissariat à l'Energie Atomique, Département de Radiobiologie et Radiopathologie, BP6, 92265-Fontenay aux Roses, France

Bootsma, D. (181)
Dept. of Cell Biology and Genetics, MGC, Erasmus University, PO Box 1738, 3000 DR Rotterdam, The Netherlands

Drabek, R. (145)
Department of Radiological Health Sciences (RHS-MRB), Colorado State University, Fort Collins, CO 80523, USA

Friedberg, E. (181)
Dept. of Pathology, University of Texas Southwestern Medical Center, Dallas, Texas, USA

Fuks, Z. (95)
Department of Radiation Oncology, Memorial Sloan-Kettering Cancer Center, New York, USA

Gamage, R. (3)
Department of Bioregulation Research, NIRS, Chiba 263, Japan

Gustafson, D. (145)
Department of Radiological Health Sciences (RHS-MRB), Colorado State University, Fort Collins, CO 80523, USA

Hama-Inaba, H. (3, 77)
Department of Bioregulation Research, NIRS, Chiba 263, Japan

Harada, Y. (175)
Genome Research Group, NIRS, Chiba 263, Japan

Higashi, T. (115)
Genome Research Group, NIRS, Chiba 263, Japan

Hirota, S. (175)
Department of Pathology, Osaka University Medical School, Osaka 565, Japan

Hoeijmakers, J.H.J. (181)
Dept. of Cell Biology and Genetics, MGC, Erasmus University, PO Box 1738, 3000 DR Rotterdam, The Netherlands

Hongo, E. (115)
Genome Research Group, NIRS, Chiba 263, Japan

Hori, T. (135)
Genome Research Group, NIRS, Chiba 263, Japan

Ichikawa, K.(13)
Faculty of Pharmaceutical Sciences, Kyushu University, Higashi-ku, Fukuoka 812-82, Japan

Ichimura, S. (49)
Division of Biology and Oncology, NIRS, Chiba 263, Japan

Ikawa, M. (175)
Research Institute for Microbial Diseases, Osaka University, Osaka 565, Japan

Ikota, N. (3, 77)
Department of Bioregulation Research, NIRS, Chiba 263, Japan

Imai, T. (135)
Genome Research Group, NIRS, Chiba 263, Japan

Inanami, O. (23)
Laboratoty of Radiation Biology, Graduate School of Veterinary Medicine, Hokkaido University, Sapporo 060, Japan

Inoue, H. (115)
Genome Research Group, NIRS, Chiba 263, Japan

Ishihara, H. (57)
First Research Group, NIRS, Chiba 263, Japan

Ishihara, Y. (115)
Genome Research Group, NIRS, Chiba 263, Japan

Ito, H. (135)
Genome Research Group, NIRS, Chiba 263, Japan

Jeggo, P.A. (125)
MRC Cell Mutation Unit, University of Sussex, Brighton, BN1 9RR, UK

Kerr, J. (69)
Formerly Department of Pathology, University of Queensland Medical School, Herston 4006, Australia

Kimura, T. (115)
Genome Research Group, NIRS, Chiba 263, Japan

Kitamura, Y. (175)
Department of Pathology, Osaka University Medical School, Osaka 565, Japan

Koike, M. (175)
Genome Research Group, NIRS, Chiba 263, Japan

Kuwabara, M. (23)
Laboratory of Radiation Biology, Graduate School of Veterinary Medicine, Hokkaido University, Sappporo 060, Japan

Masumizu, T. (13)
ESR Application Laboratory, Application and Research Center, JEOL LTD., Akishima, Tokyo 196, Japan

Matsuda, Y. (135)
Genome Research Group, NIRS, Chiba 263, Japan

McGraw, M. (145)
Department of Radiological Health Sciences (RHS-MRB), Colorado State University, Fort Collins, CO 80523, USA

Mita, K. (49, 115)
Genome Research Group, NIRS, Chiba 263, Japan

Miura, Y. (13)
Department of Bioregulation Research, NIRS, Chiba 263, Japan

Morimyo, M. (115)
Genome Research Group, NIRS, Chiba 263, Japan

Nagase, T. (135)
Laboratory of Gene Structure I, Kazusa DNA Research Institute, Chiba 292, Japan

Nenoi, M. (49)
Division of Biology and Oncology, NIRS, Chiba 263, Japan

Nohata, J. (115)
Genome Research Group, NIRS, Chiba 263, Japan

Nomura, N. (135)
Laboratory of Gene Structure I, Kazusa DNA Research Institute, Chiba 292, Japan

Ohyama, H. (77)
Div. of Radiobiol. and Biodosimetry, NIRS, Chiba 263, Japan

Okabe, M. (175)
Research Institute for Microbial Diseases, Osaka University, Osaka 565, Japan

Ozawa, T. (3)
Department of Bioregulation Research, NIRS, Chiba 263, Japan

Packer, L. (33)
Department of Molecular and Cell Biology, University of California, Berkeley, CA, USA

Park, W.Y. (115)
Chungbuk National University, Korea

Robinson, J. (145)
Department of Radiological Health Sciences (RHS-MRB), Colorado State University, Fort Collins, CO 80523, USA

Sagara, M. (135)
Genome Research Group, NIRS, Chiba 263, Japan

Saito, T. (135)
Genome Research Group, NIRS, Chiba 263, Japan

Sano, H. (13)
Faculty of Pharmaceutical Sciences, Kyushu University, Higashi-ku, Fukuoka 812-82, Japan

Sasanuma, S. (115)
Genome Research Group, NIRS, Chiba 263, Japan

Satoh, T. (3)
Tokushima Bunri University, Tokushima 770, Japan

Seki, N. (135)
Laboratory of Gene Structure I, Kazusa DNA Research Institute, Chiba 292, Japan

Shimazu, Y. (3)
Department of Bioregulation Research, NIRS, Chiba 263, Japan

Shimokawa, T. (77)
Div. of Radiobiol. and Biodosimetry, NIRS, Toho Univ. Graduate Sch. Biomol. Sci., Biochem. Div., Natl. Cancer Ctr. Res. Inst., Dept. Radiol. Health, Faculty of Med., Univ. of Tokyo, Tokyo, Japan

Shiomi, N. (175)
Genome Research Group, NIRS, Chiba 263, Japan

Shiomi, T. (175)
Genome Research Group, NIRS, Chiba 263, Japan

Strauss, B.S. (151)
Department of Molecular Genetics and Cell Biology, The University of Chicago, Illinois 60637, USA

Sugawara, T. (135)
Genome Research Group, NIRS, Chiba 263, Japan

Sugaya, K. (115)
Genome Research Group, NIRS, Chiba 263, Japan

Takeshita, K. (13)
Faculty of Pharmaceutical Sciences, Kyushu University, Higashi-ku, Fukuoka 812-82, Japan

Takusakawa, M. (3)
Department of Bioregulation Research, NIRS, Chiba 263, Japan

Tanaka, I. (57)
First Research Group, NIRS, Chiba 263, Japan

Tanaka, K. (165, 181)
Institute for Molecular and Cellular Biology, Osaka University, Osaka 565, Japan

Tanuma, S. (85)
Department of Biochemistry, Faculty of Pharmaceutical Sciences, Science University of Tokyo, Shinjuku-ku, Tokyo, Japan

Tsuji, S. (115)
Genome Research Group, NIRS, Chiba 263, Japan

Tsuneoka, K. (57)
First Research Group, NIRS, Chiba 263, Japan

Ueda, J. (3)
Department of Bioregulation Research, NIRS, Chiba 263, Japan

Ueno, A. (145)
Department of Radiological Health Sciences (RHS-MRB), Colorado State University, Fort Collins, CO 80523, USA

Utsumi, H. (13)
Faculty of Pharmaceutical Sciences, Kyushu University, Higashi-ku, Fukuoka 812-82, Japan

van der Horst, G.T.J. (181)
Dept. of Cell Biology and Genetics, MGC, Erasmus University, PO Box 1738, 3000 DR Rotterdam, The Netherlands

Vannais, D. (145)
Department of Radiological Health Sciences (RHS-MRB), Colorado State University, Fort Collins, CO 80523, USA

van Steeg, H. (181)
Dept. LCM, RIVM, Bilthoven, The Netherlands

Waldren, C. (145)
Department of Radiological Health Sciences (RHS-MRB), Colorado State University, Fort Collins, CO 80523, USA

Yamada, T. (77)
Toho Univ. Graduate Sch. Biomol. Sci., Tokyo, Japan

Yamauchi, M. (115, 135)
Genome Research Group, NIRS, Chiba 263, Japan

Yukawa, O. (77)
Div. of Biol. and Oncol., NIRS, Chiba 263, Japan

Contents

III Apoptosis as a Biodefence Mechanism

IV Radiation-induced DNA Damages, Repair and Mutagenesis

V Repair of UV-induced DNA Damages

Preface

Since life was first spawned on earth, all living cells have been engaged in the development of a network of biodefence systems to protect against the deleterious effects of various environmental stress. Besides exposures to so-called pollutants, human beings breathe oxygen and live in a world of unavoidable exposure to endogenous agents such as active oxygens and other free radicals, ionizing radiation, ultraviolet light and chemicals from such natural sources as normal metabolism. To minimize the harmful effects of these exposures, it is necessary to improve our understanding of fundamental life processes that preserve the genome.

A new and better understanding of biodefence mechanisms will contribute to practical fields such as risk assessment, identification of high risk groups and individuals, and the prevention and treatment of genetic diseases such as cancer caused by environmental stress. Besides direct effects on DNA, active oxygens and free radicals are believed to cause functional defects in living organisms by affecting signal transduction associated with the activation of specific genes, which can in turn lead to cancer and premature aging.

We are, however, beginning to unravel the complexity and understand the intricate details of the surveillance and repair systems that identify and repair damaged DNA. Progress has been made possible by the identification, cloning and characterization of many of the genes that participate in various repair pathways in humans and in other informative organisms such as bacteria, yeast and rodents. Much of this information has come from genome research projects carried out in these organisms.

In view of this, the National Institute of Rediological Sciences (NIRS) organized the "International Symposium on Biodefence Mechanisms: Search for Novel Functions in Biodefence against Environmental Stress" to promote the exchange of state-of-the art information among specialists working in various life sciences and to delineate a perspective for future research on biodefence mechanisms. The symposium focused on five related topics: 1) protection against oxidative stress-induced damages, 2) stress-induced signal transduction systems for biodefence, 3) apoptosis as a biodefence mechanism, 4) radiation-induced DNA damages, repair and mutagenesis, and 5) repair of UV-induced DNA damages.

The Proceedings include three Special Lectures and sixteen papers from the symposium. It is our hope that this volume will contribute to the advancement of science and technology of biodefence against environmental stress. We are confident that fundamental research of this kind will provide information that will help prevent and treat diseases caused by environmental stress.

On behalf of the organizing committee, we would like to express our heartfelt thanks to Japan International Science and Technology Exchange Center (JISTEC) for its co-sponsorship and also to the members of the program committee, chairpersons, panelists, invited speakers, and all other participants.

October 1997

Toshihiko Ozawa
Kouichi Tatsumi
Tada-aki Hori

I

Protection against Oxidative Stress-induced Damages

1. Histidine-containing Peptides and Hydroquinone Derivatives as Synthetic Antioxidants

Nobuo IKOTA, Jun-ichi UEDA, Ranjith GAMAGE, Yoshie SHIMAZU, Hiroko HAMA-INABA, Mitsuko TAKUSAKAWA, Toshihiko OZAWA, and Toshio SATOH*

Department of Bioregulation Research, NIRS, Chiba 263, Japan and *Tokushima Bunri University, Tokushima 770, Japan

Active oxygen species have been implicated in the aging process and in a wide variety of diseases such as cancer and arthritis. Radiation induced tissue damage including damage from ultraviolet exposure is also a direct result of free radical-mediated toxicity. Therefore the prevention of oxidative damage is a subject of considerable investigation. Among the most prevalent reactive oxygen species in biological systems (O_3(ozone), $^3O_2$(molecular oxygen), $^1O_2$(singlet oxygen), O_2^- (superoxide), H_2O_2(hydrogen peroxide), OH·(hydroxyl radical), RO·(alkoxyl radical), ROO· (alkylperoxy radical), epoxide), the hydroxyl radical is extremely reactive and responsible for cellular damage such as protein oxidation, lipid peroxidation, and DNA breakage.

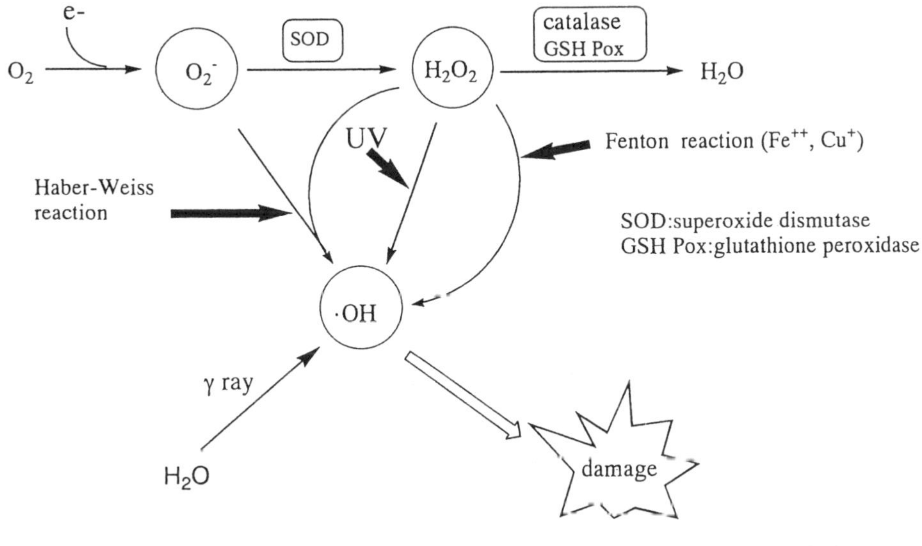

Fig.1 Generation and removal of active oxygen species (O_2^-, · OH, H_2O_2)

As shown in Fig. 1, one electron reduction of molecular oxygen generates O_2^-, and then O_2^- is dismutated to O_2 and H_2O_2 by superoxide dismutase (SOD). H_2O_2 thus formed is converted to O_2 and H_2O by catalase or glutathion peroxidase. However, if metals such as Cu^+ and Fe^{++} are present, hydroxyl radical is generated from H_2O_2 by Fenton reaction. The hydroxyl radical can also be generated by Haber-Weiss reaction from H_2O_2 and, UV irradiation of H_2O_2, and γ-ray irradiation of H_2O. Therefore, in the present study, the generation and removal of O_2^- and hydroxyl radical using synthetic compounds (Fig. 2, Cu-complexes of peptides containing histidyl residue[1]) (HisGly (**1**), HisGlyGly (**2**), HisGlyGlyGly (**3**), GlyHisGly (**4**), GlyGlyHisGly (**5**)) , Histidine-containing peptides[2] (Ac-D-PenGlyHisGly (Ac-D-Pen-GHG, **6**), MetGlnPheHisThr (MEFHT,**7**), ProHisCysLysArgMet (PHCKRM, **8**)), and hydroquinone derivatives(4-hexyloxy-2,3,6-trimethyl-phenol(**9**)[3] and (2*S*,3*R*,4*S*)-*N*-(3-(1-hydroxy-2,3,6-trimethylphenoxy)propyl)-3,4-dihydroxy-2-hydroxymethylpyrrolidine (**10**)), and hydroxychalcon (2',4'-3,4-tetrahydroxychalcone (**11**)[4]) are discussed. Peptides **7** and **8** were the substructures of the protein, which was reported as the resistant protein against active oxygen species.[5]

His-Gly(1) His-Gly-Gly(2) His-Gly-Gly-Gly(3)

Gly-His-Gly(4) Gly-Gly-His-Gly(5)

Ac-D-Pen-Gly-His-Gly (6, Ac-D-Pen-GHG)

Pro-His-Cys-Lys-Arg-Met (7, PHCKRM)

Met-Gln-Phe-His-Thr (8, MEFHT)

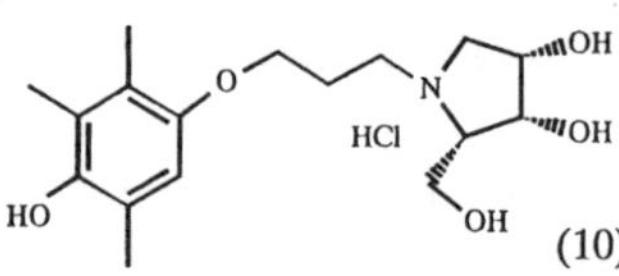

(10)

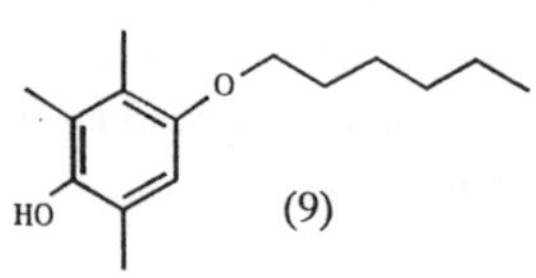

(9)

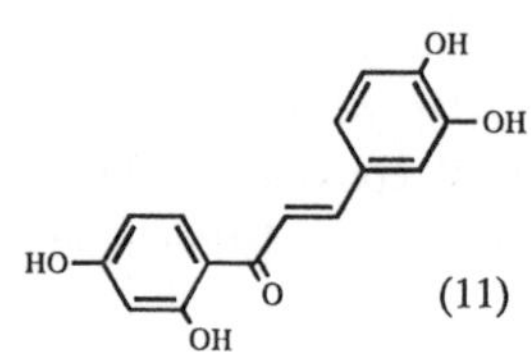

(11)

Fig. 2 Synthetic compounds

Histidine-containing di,tri and tetra-peptides were synthesized by conventional liquid phase method. Protected amino acids derivatives were condensed stepwise from glycine. The synthesis of Ac-D-Pen-GHG (**6**) is described as representative. Glycine benzyl ester *p*-toluenesufonic acid salt was condensed with Boc-*p*-methoxybenezenesulfonyl histidine using diethyl phosphorocyanidate (DEPC)[6] as a coupling reagent to afford the protected di-peptide. After removal of Boc group by trifluoroacetic acid, Boc-glycine was coupled again using DEPC, and Boc group of the resulting tri-peptide was removed by trifluoroacetic acid followed by condensation with Ac-D-Penicillamin using DEPC to afford the protected tetra-peptide. All protecting groups were cleaved with anhydrous hydrogen fluoride in the presence of anisole to give the free peptide after purification by ion-exchange chromatography and reprecipitation from H_2O-EtOH. The two peptides (**7** and **8**) were synthesized by solid phase method using fluorenylmethoxycarbonyl (Fmoc) methodology. Removal of all the protecting groups and cleavage from polymer were performed with trifluoroacetic acid and crude material was purified by HPLC. 4-Hexyloxy-trimethylhydroquinone (**9**) was synthesized from trimethylhydroquinone and hexanol in the presence of phosphomolybdic acid. A water soluble derivative of hydroquinone mono-alkylether (**10**) was prepared from 3-(1-(*tert*-butyldimethylsilyloxy)-2,3,6-trimethylphenoxy)propylbromide and (2*S*,3*R*,4*S*)-3,4-isopropylidenedioxy-2-trityl-oxymethyl pyrrolidine[7] in the presence of potassium carbonate in acetone followed by removal of all protecting groups by acidic hydrolysis (10% HCl-MeOH). In this compound, the polyhydroxylated pyrrolidine moiety was expected to be the hydrophilic part. Hydroxychalcone (**11**) was synthesized from 2,4-dihydroxyacetophenone and 3,4-dihydroxybenzaldehyde.

Since O_2^- and hydroxyl radical are not stable, the generation and removal of active oxygen species were investigated by ESR spin trapping method using DMPO(5,5-dimethyl-1-pyrroline *N*-oxide). First, the reactivities of Cu(II) complexes of the histidine-containing peptides (**1**-**5**) toward O_2^- were investigated by ESR spin trapping method. ESR peak height ratio of DMPO-O_2^- spin adduct and reference sample(MnO) was plotted against the concentration of the copper complex, then IC_{50} was calculated. As shown in Table 1, all Cu(II) complexes examined could scavenge O_2^-, although their scavenging abilities are not so high as SOD itself. Among these peptides, Cu-complexes of peptides containing histidyl residue at *N*-terminal (**1**, **2** and **3**) showed comparatively higher reactivity toward O_2^- than those with histidyl residue at second (**4**) or third position (**5**).

Table 1 Ability of Cu(II)-peptide complexes for dismutation of superoxide

copper complexes	IC50
Cu,Zn-SOD	0.008 μM
Cu(II)-HisGly	1.3
Cu(II)-HisGlyGly	0.8
Cu(II)-HisGlyGlyGly	1.1
Cu(II)-GlyHisGly	7.5
Cu(II)-GlyGlyHisGly	19.0

Cu(II) peptide complexes were prepared from peptides (1.1 eq) and copper sulfate
Concentrations: HPX, 0.5mM; XOD, 0.1 unit/ml; DMPO, 125 mM.

During the experiment on superoxide dismutation using Cu(II) complex of cyclohistidine, which is known as the SOD model,[8] we found a high concentration of this Cu(cyclohistidine)$_2$ about 0.3 mM generated the hydroxyl radical with complete disappearance of O_2^-(Fig. 3). At a Cu-complex concentration of around 0.03 mM, the removal of O_2^- was complete and at the same time the formation of hydroxyl radical was observed, and gradually increased with increasing concentration of Cu-complex. The appearance of low signal intensity of DMPO-OH adduct in the region of low Cu concentration was explained by the reaction of unreacted O_2^- and H_2O_2 produced from O_2^- with SOD activity by Haber-Weiss reaction. On the other hand, at the high concentration of Cu-complex, O_2^- was entirely dismutated to O_2 and H_2O_2 by Cu complex, and H_2O_2 thus formed can react with Cu complex to generate hydroxyl radical by Fenton type reaction.

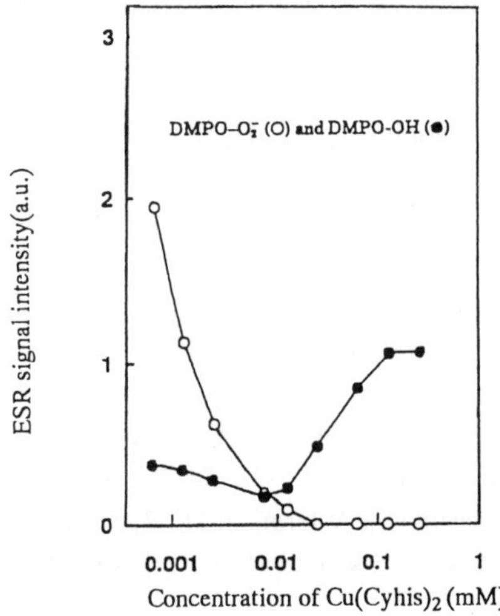

Reaction mixtures contained 0.5 mM HPX, 0.1 unit/ml XOD, and 125 mM DMPO

Fig. 3 Effectt of concentration of $Cu(Cyhis)_2$ on superoxide dismutation

From these experiments, we next examined the generation of hydroxyl radical from Cu complexes of these histidine-containing peptides with H_2O_2. The results are summarized in Table 2. We have already reported the generation of OH radical from the reaction of $Cu(en)_2$ complex with H_2O_2.[9] Therefore, the values indicated in Table 2 are relative to that obtained from the reaction of $Cu(en)_2$ and H_2O_2. The strong signal intensity of DMPO-OH was obtained from the reaction of H_2O_2 with Cu complex of peptides containing histidyl residue at *N*-terminal, whereas the weak signal intensity was observed from Cu complex of peptides containing histidyl residue in the second or third position. Hydroxyl radical could be generated by the reactions of Cu(II) complexes with H_2O_2 in sequential equations as shown in Scheme 1.

Table 2 Formation of DMPO-OH from the reaction of $Cu(II)(en)_2$ and Cu(II)-histidine oligopeptide complexes with H_2O_2

Copper(II) Complexes	DMPO-OH (relative intensity)
$Cu(en)_2$	100
Cu(II)-HisGly	115
Cu(II)-HisGlyGly	118
Cu(II)-HisGlyGlyGly	150
Cu(II)-GlyHisGly	5
Cu(II)-GlyGlyHisGly	27

concentration: Cu complex; 0.25 mM, H_2O_2; 25 mM, DMPO; 25 mM

$$Cu(II) + H_2O_2 \longrightarrow Cu(I) + O_2^- + 2H^+$$

$$Cu(II) + O_2^- \longrightarrow Cu(I) + O2$$

$$Cu(I) + H_2O_2 \longrightarrow Cu(II) + \cdot OH + OH^-$$

Scheme 1 Assumed reaction sequence of Cu(II)-peptide complex with H_2O_2

There has been recent interest in the reaction of DNA with active oxygen species. Therefore, we examined DNA damage caused by Cu complexes of histydine-containing peptides in the presence of H_2O_2 (Fig. 4).[10] In the treatment of DNA with H_2O_2 and with $Cu(en)_2$, the migration

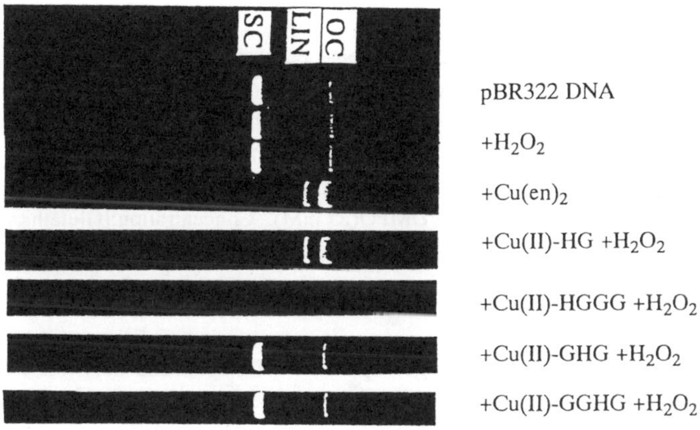

Fig. 4 Reaction of supercoiled DNA with Cu(II) complex in the presence of H_2O_2

patern did not change, indicating these agents did not cause DNA strand scission. On the other hand, DNA breakage to the open circular and linear form was observed from the reaction of $Cu(en)_2$ with H_2O_2. When Cu complex of peptide with histidine at *N*-terminal and H_2O_2 were used, DNA breakage was also observed. In particular, Cu-HGGG caused more extensive DNA strand break; as a result, no open circular or linear forms were observed. In contrast, DNA strand scission was not observed in the reaction of the Cu complexes of the peptides with histidine in the second or third position (Cu-GHG and Cu-GGHG). There is a good correlation between DMPO-OH formation by ESR experiments and the extent of DNA strand scission. Therefore, it is concluded that DNA strand scission was caused by the hydroxyl radical.

Next, we examined the scavenging abilities of histidine-containing peptides (**6**,**7**, and **8**) against O_2^- and hydroxyl radical by ESR spin trapping method. Their scavenging abilities were estimated by the reduction of ESR signal intensity of DMPO-adduct. Table 3 shows the scavenging abilities against O_2^- and hydroxyl radical at the concentration of 0.25 mmol. Ac-D-Pen-GHG and MEFTH have almost no scavenging abilities against O_2^-. PHCKRM has a weak scavenging ability, almost the same as glutathion. As for the scavenging ability against hydroxyl radical generated from $Cu(en)_2$ and H_2O_2, these two peptides have higher scavenging abilities than glutathione. Particularly, PHCKRM showed the highest scavenging ability. These scavenging

Table 3 Scavenging ability of oligopeptides 6, 7, and 8 against O_2^- and hydroxyl radical

Compound	% inhibition of DMPO-O_2^-*	% inhibition of DMPO-OH**
None	0	0
Cu,Zn-SOD	100	-
Glutathione	31	64
Ac-D-Pen-Gly-His-Gly	0	85
Met-Gln-Phe-His-Thr	2	0
Pro-His-Cys-Lys-Arg-Met	25	96

* O_2^- was generated by the HPX(0.5 mM)/XOD(0.1 U/ml) system at pH 7.4 in 0.1M phosphate buffer in the presence of DMPO(125 mM). Concentration: [Cu,Zn-SOD]=0.25 mg/ml ; [Glutathione]=[Met-Gln-Phe-His-Thr]=[Ac-D-Pen-Gly-His-Gly]= [Pro-His-Cys-Lys-Arg-Met]= 0.25 mM

** OH radical was generated from the reaction of $Cu(en)_2$(0.25 mM) with H_2O_2(25 mM) at pH 7.4 in 0.1 M phosphate buffer in the presence of DMPO(25 mM). Concentration:[Glutathione]=[Met-Gln-Phe-His-Thr]=[Ac-D-Pen-Gly-His-Gly]= [Pro-His-Cys-Lys-Arg-Met]=0.25 mM

abilities were confirmed by the DNA breakage experiment. DNA strand scission was not observed from the reaction of Cu(en)2 with H2O2 in the presence of two peptides(**6** and **8**) as shown in Fig. 5.

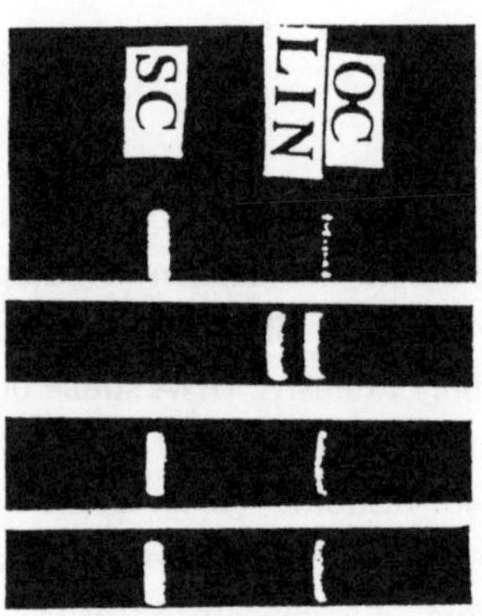

pBR322 DNA

+$Cu(en)_2$ +H_2O_2

+$Cu(en)_2$ +H_2O_2 + Ac-D-Pen-GHG

+$Cu(en)_2$ +H_2O_2 + PHCKRM

Fig. 5 Reaction of supercoiled DNA with $Cu(en)_2$ and oligopeptides (Ac-D-Pen-GHG and PHCKRM) in the presence of H_2O_2

There are two mechanisms for the suppression of DMPO-OH adduct formation by these peptides. First, direct scavenging for OH radical, and second suppression of the generation of hydroxyl radical. In the second case, the peptide may bind to the metal, and the metal-peptide complexes cannot activate H2O2 to generate the hydroxy radical. In order to determine the

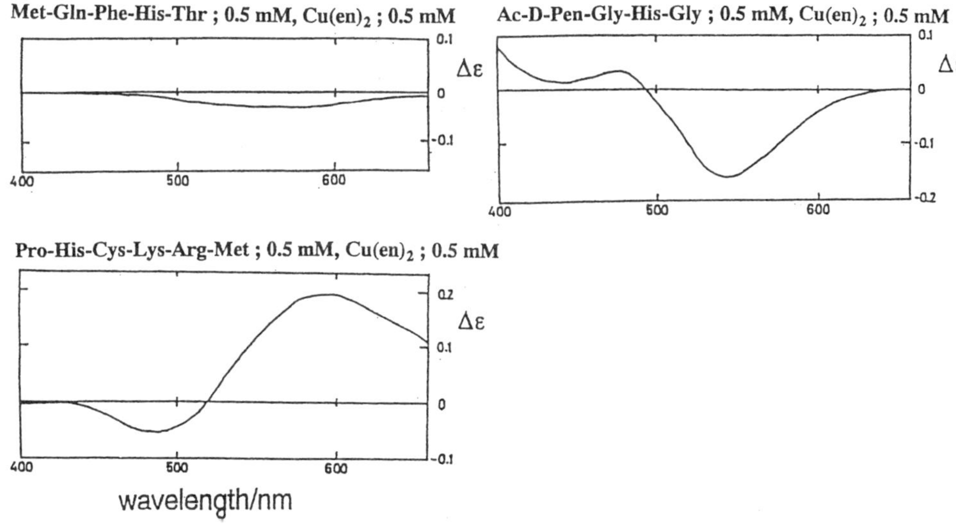

Fig. 6 CD spectra of reaction mixture of $Cu(en)_2$ with oligopeptides 6, 7, and 8

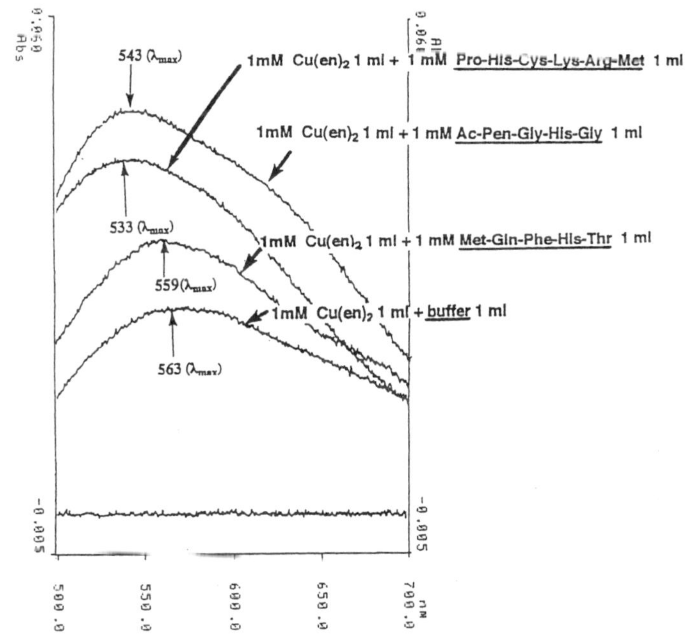

Fig. 7 Visible absorption spectra of reaction mixture of $Cu(en)_2$ with oligopeptides 6, 7, and 8

mechanism, reaction mixture of $Cu(en)_2$ and the peptides **6**, 7, and **8** were examined by CD and visible absorption spectra. Since ethylenediamine does not possess a chiral center, $Cu(en)_2$ complex has no CD absorption. If a ligand with a chiral center binds to Cu ion, this complex should have a CD absorption. As shown in Fig. 6, the reaction mixture of $Cu(en)_2$ with Ac-D-Pen-GHG and

PHCKRM showed strong CD absorption, whereas the reaction mixture of $Cu(en)_2$ with MEFTH has only weak absorption. In the visible absorption spectra (Fig. 7), λ_{max} for $Cu(en)_2$ solution and λ_{max} for the reaction mixture of $Cu(en)_2$ and MEFTH were almost same(563 nm and 559 nm, respectively), while in the mixture of $Cu(en)_2$ and peptides **6** and **8**, λ_{max} (543 nm and 533 nm) were shifted more than 20 nm from that of $Cu(en)_2$ solution. These results suggest that these two peptides can coordinate strongly to the Cu ion .

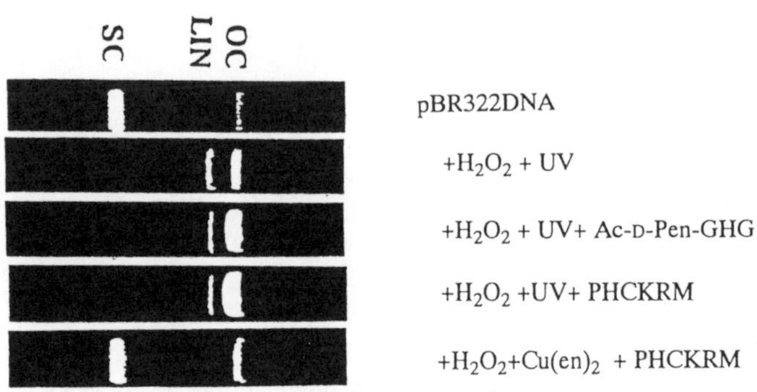

Fig. 8 Reaction of supercoiled DNA with ·OH generated by H_2O_2 and UV irradiation and oligopeptides (Ac-D-Pen-GHG and PHCKRM)

Peptides **6** and **8** were then examined to see whether they scavenge the hydroxyl radical directly or not by DNA breakage experiment (Fig. 8). In the reaction of DNA at UV irradiation of H_2O_2, breakage occurred. In this system, Ac-D-Pen-GHG and PHCKRM were added. In these cases, DNA breakages were observed (compared with experiments using $Cu(en)_2$ and H_2O_2), although these peptides slightly scavenged hydroxyl radical. Therefore, the direct scavenging ability of these peptide was not so high. In conclusion, two peptides (**6** and **8**), which have histidyl residue in the second and third position, inhibited the generation of hydroxyl radical in the reaction of $Cu(en)_2$ and H_2O_2 by complexation with Cu(II) ion.

Next, the scavenging abilities of compounds **9**, **10** and **11**toward O_2^- and hydroxyl radical using ESR spin trapping method were examined and the results are listed in Tables 4 and 5. As shown in Table 4, compounds **9** and **11** did not dissolve in phosphate buffer. Therefore, they were measured in dimethyl sulfoxide solution. Hydroquinone derivatives (**9** and **10**) showed only weak scavenging ability against O_2^-. On the other hand, hydroxychalcone (**11**) showed very high scavenging ability toward O_2^-. Table 5 shows the results of scavenging ability against OH radical generated from $Cu(en)_2$ and H_2O_2. The compounds **9** and **11** were measured in 50%EtOH solution due to their insolubility in phosphate buffer. Therefore, the spin-adduct was not DMPO-OH but DMPO-1-hydroxyethyl, because hydroxyl radical reacts rapidly with EtOH to generate the 1-hydroxyethyl radical. The compounds **9**, **10**, and **11** showed comparatively high scavenging ability against hydroxyl radical. Hydroxychalcone **11** especially showed high scavenging ability toward both hydroxyl radical and O_2^-.[11] Very interestingly, radiation-induced apoptotic DNA fragmentation and increase of peroxide were inhibited by compound **11**, added before or immediately after irradiation.[12]

2. *In Vivo* ESR Measurement of Free Radical Reactions in Living Mice

Hideo UTSUMI*, Keizo TAKESHITA, Kazuhiro ICHIKAWA,
Hiroaki SANO, Toshiki MASUMIZU[1], and Yuri MIURA[2]

Faculty of Pharmaceutical Sciences, Kyushu University,
Higashi-ku, Fukuoka 812-82, Japan

Introduction

Free radicals such as active oxygen species and nitric oxide are believed to be functional compounds in various biological systems, and numerous reports have been made to clarify the roles of free radicals in biological phenomena. However, most reports are of *in vitro* experiments and there are few *in vivo* investigations. Non-invasive measurement of *in vivo* free radical reactions is very important to understand the biological function of free radicals, since oxygen concentration in tissues of living body is much lower than that in *in vitro* experiments and there are many substances and reaction pathways which influence *in vivo* radical reaction.

The recent development of a low-frequent ESR spectroscopy has enabled non-invasive *in vivo* measurement of radicals in whole animals.[1-7] ESR spectra of nitroxyl radicals are susceptible to oxygen concentration,[8] to active oxygen species[9] and to biological redox systems,[10] indicating that nitroxyl radicals may provide, as a probe of *in vivo* ESR measurement, much valuable information on the biological functions of free radical reactions including generation of active oxygen species and activity of redox systems. Using the *in vivo* ESR-CT system we have studied the spectral change in nitroxyl radicals after intravenous (i.v.),[6, 11-14] intramuscular (i.m.),[12-15] intraperitoneal (i.p.), [16] and intratracheal (i.t.) [17, 18] injection to mice. The results indicate reduction of nitroxyl radicals and reveal that the rate of reduction depends on physiological and pathological conditions such as aging, oxidative stress, etc.

Footnote ;
* To whom all correspondence should be addressed.
Present address; 1) ESR Application Laboratory, Application and Research Center, JEOL LTD., Akishima, Tokyo 196 and 2) Department of Bioregulation Research. NIRS, Chiba 263, Japan

In this chapter, we review our results about non-invasive ESR measurement of *in vivo* free radical reactions in living mice using nitroxyl radicals as probes and discuss the utilization of this technique for clarifying their role in biological response.

Materials and Methods

Table 1 demonstrates the major nitroxyl probes used in our laboratory. Most 2,2,6,6-tetramethylpiperidine-1-oxyl (TEMPO) and 2,2,5,5-tetramethylpyrrolidine-1-oxyl (PROXYL) derivatives were purchased from Aldrich Chemical Co.. Glutaramide- and methylglutaramide-TEMPO and 4,4-dimethyloxazolidine-1-oxyl (OXANO) derivatives were synthesized in our laboratory. The water soluble probes were dissolved in isotonic solution, and the osmolarity was measured by freezing point depression. Pentobarbital sodium was purchased from Dainabot Co. Ltd. Other reagents were of the highest purity commercially available.

Female ddY mice (3-4 weeks old, 15-20 g body weight) were used throughout this study. Mice were anesthetized by i.m. or i.p. injection of pentobarbital (50 mg/kg). Immediatetly after

Table 1. Acronym and structures of spin-probes used for *in vivo* ESR measurement.

acronym	chemical name	R	basic structure
1.TEMPO derivatives			
TEMPO	2,2,6,6-tetramethylpiperidine-1-oxyl	-H	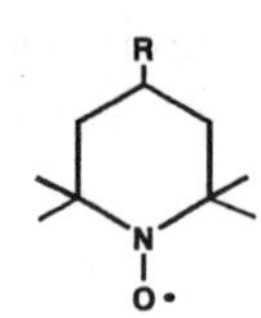
hydroxy-TEMPO	4-hydroxy-2,2,6,6-tetramethylpiperidine-1-oxyl	-OH	
amino-TEMPO	4-amino-2,2,6,6-tetramethylpiperidine-1-oxyl	$-NH_2$	
carboxy-TEMPO	4-carboxy 2,2,6,6-tetramethylpiperidine-1-oxyl	-COOH	
oxo-TEMPO	4-oxo 2,2,6,6-tetramethylpiperidine-1-oxyl	=O	
CAT-1	4-trimethylammonium-2,2,6,6-tetramethylpiperidine-1-oxyl iodide	$-N^+(CH_3)I^-$	
phosphonooxy-TEMPO	4-phosphonooxy 2,2,6,6-tetramethylpiperidine-1-oxyl	$-OPO_3H_2$	
glutarammide-TEMPO		$-NHCO(CH_2)_3COOH$	
methylglutaramide-TEMPO		$-NHCO(CH_2)_3COOCH_3$	
2.PLOXYL derivatives			
carboxy-PROXYL	3-carboxy-2,2,5,5-tetramethylpyrrolidine-1-oxyl	-COOH	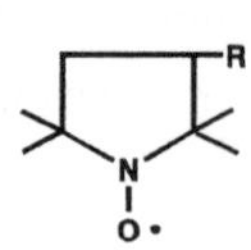
carbamoyl-PROXYL	3-carbamoyl-2,2,5,5-tetramethylpyrrolidine-1-oxyl	$-CONH_2$	
aminomethyl-PROXYL	3-aminomethyl-2,2,5,5-tetramethylpyrrolidine-1-oxyl	$-CH_2NH_2$	
3.OXANO derivatives			
2-DOXYL-butane	2-ethyl-2,4,4-trimethyloxazolidine-3-yloxy		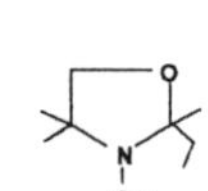
nSLS	n-(N-oxyl-4'4'-dimethyloxazolidine)-stearic acid		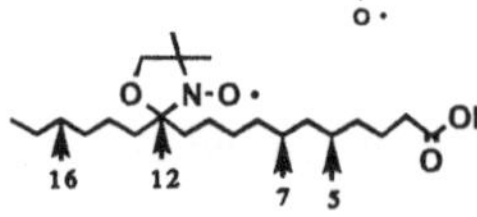

a)

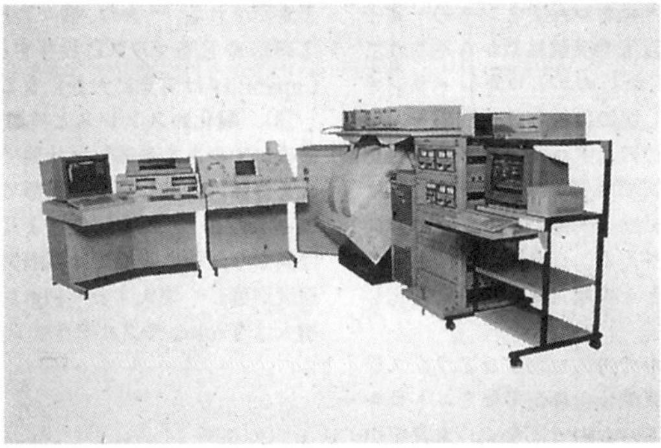

b)

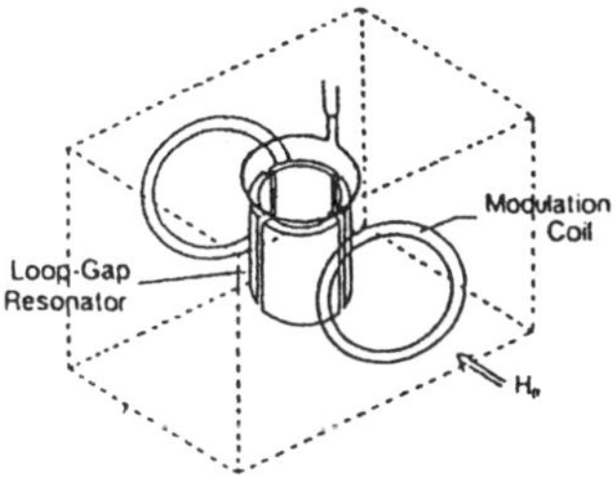

Figure 1. Picture of *in vivo* ESR-CT system (a) and structure of a loop-gap resonator b) in our laboratory.[11]

administration of isotonic solution of spin-probes through various routes, ESR spectra were measured at different domains from head to tail with an *in vivo* ESR-CT system. The microwave frequency was 1.1-1.3 GHz and the power was 1.0 - 5.0 mW. The amplitude of the 100 KHz field modulation was 0.1 - 0.2 mT. The external magnetic field was swept at a scan rate of 5 mT/min.

Results & Discussion

Characteristics of our *in vivo* ESR-CT system in our laboratory

Figures 1a and 1b show our *in vivo* ESR-CT system and the fundamental structure of a loop-gap resonator of the ESR system, respectively. The resonator has 2 or 4 gaps and its division is 33 mm i.d. and 5 or 24 mm long. The maximum amplitude of 100 kHz field modulation is *ca.* 0.2 mT. The sensitivity of the *in vivo* ESR apparatus is about 1/40 - 1/300 of that of a conventional X-band ESR spectrometer, depending on the weight of animals and the observed domain. The spatial distribution of the sensitivity was fairly homogeneous in the resonator.[11]

ESR spectra of Nitroxyl Radicals in Whole Mice

Figures 2a and 2b show typical ESR spectra at the hepatic domain of a mouse with carbamoyl-PROXYL and hydroxy-TEMPO administered into the tail vein, respectively.[6] Three sharp lines were observed with regular noise due to respiration. The hyperfine structure and the peak height ratios of carbamoyl-PROXYL coincided with those of the probe dissolved in saline at a concentration of less than 2 mM, suggesting that the spin-probe should exist as a free monomer at the hepatic domain. Hydroxy-TEMPO in the hepatic domain of mice also gave triplet lines, but the intensity of the lines decreased gradually during field sweep. Quite similar spectra were also observed at head, chest, and lower abdomen after intravenous administration.[6, 13] Figure 3 is the ESR-CT imaging picture of carbamoyl-PROXYL at abdomen of mouse after administration to the tail vein, indicating the existence of carbamoyl-PROXYL in a bundle of the inferior vena cava, hepatic artery, and celiac aorta of mice after i.v. injection.[19]

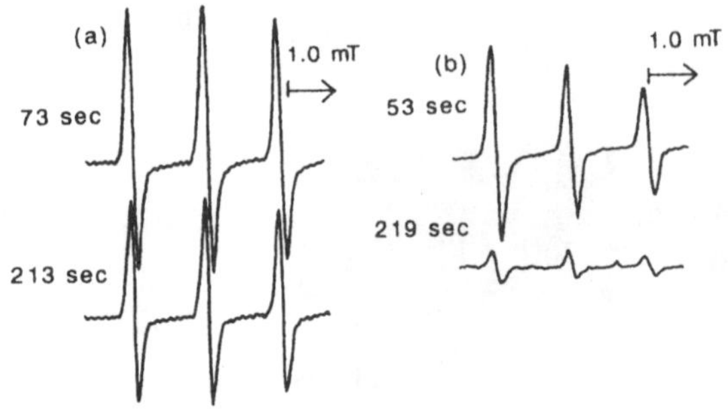

Figure 2. ESR spectra of carbamoyl-PROXYL (a) and hydroxy-TEMPO (b) at hepatic domain of mouse. One hundred microliters of an isotonic solution of spin-probes (280 mosM) was injected into the tail vein of a female ddY mouse and the ESR spectrum at hepatic domain of mouse was observed with *in vivo* ESR spectrometer. The number at the left of each spectrum indicates the time after injection.[6)]

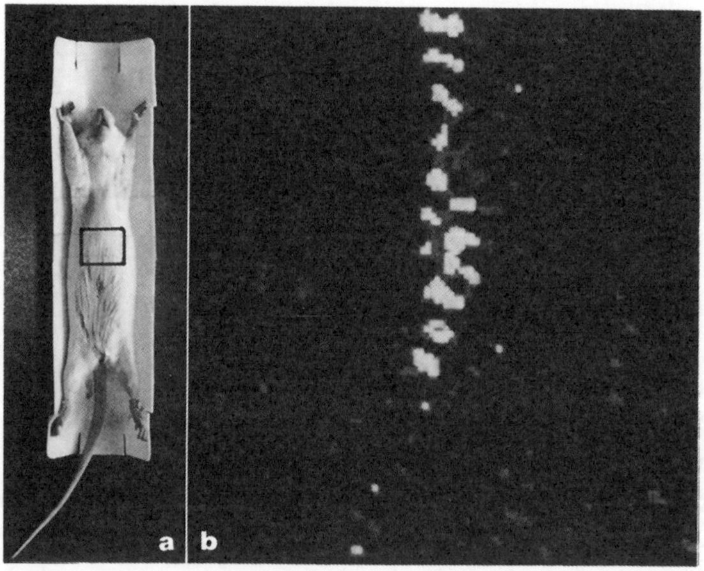

Figure 3. ESR image of nitroxyl-probe at upper abdomen of mouse after injection of carbamoyl-PROXYL (280 mosM) into the tail vein.[19)] a) Mouse used for ESR imaging. b) Image from ventral side.

I.p. or i.m. administration of spin-probes gave also the same spectra as shown in Figure 2, but the period of signal appearance depended on the administration route.[15, 16] Spin-probes injected i.t. into mice lung also showed the same triplet lines as those in water.[17, 18] Figure 4 demonstrates the ESR-CT imaging of carboxy-PROXYL injected into mouse lung. The scale and position of the image agree well with those of the lung, suggesting that the spin-probe is distributed throughout the lung.[17]

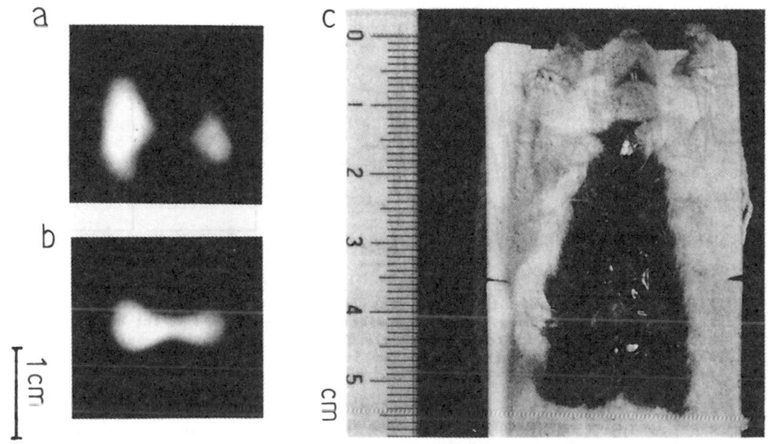

Figure 4. ESR image of nitroxyl-probe injected into mouse lung (a, b) and an anatomical picture of the mouse used for ESR image (c). ESR image was performed from two directions after transtracheal injection of carboxy-PROXYL (15 mM, 0.9 ml) to mouse lung; from the ventral side (a) and from the tail along the body axis (b).[17]

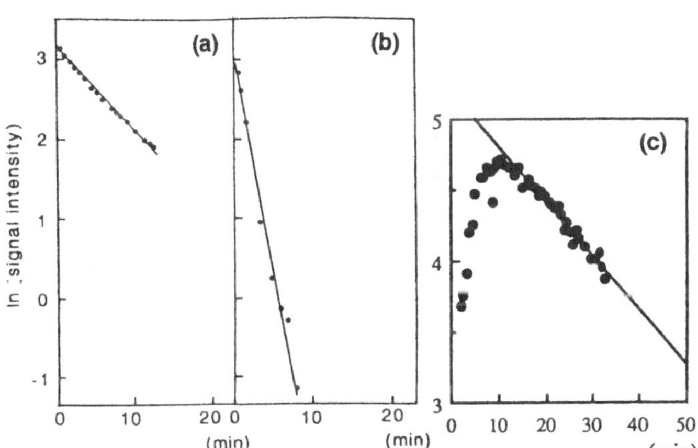

Figure 5. Decay curve of carbamoyl-PROXYL (a) and hydroxy-TEMPO (b) at hepatic domain of mouse after i.v. injection and that of carbamoyl-PROXYL (c) at head after intraperitoneal injection.[6, 16]

Signal Reduction of Nitroxyl Radicals in Living Mice

Nitroxyl radicals are readily reduced to their corresponding hydroxylamines through enzymatic and non-enzymatic processes, resulting in the loss of paramagnetism.[20] In fact, *in vivo* ESR signal of nitroxyl radicals decreased gradually after injection to living mice. Figures 5a and 5b show semilogarithmic plot of the peak heights at hepatic domain after intravenous administration of carbamoyl-PROXYL and hydroxy-TEMPO, respectively.[6] The plot was a straight line until at least 10 min, indicating that signal decay after i.v. injection should obey first order kinetics. Figure 5c shows a decay curve of carbamoyl-PROXYL at head after i.p. administration.[16] The i.p. injection of nitroxyl probes required *ca.* 10 min to reach the maximum peak level, while it reached maximum immediately after i.v. injection.

Table 2 demonstrates the *in vivo* decay constants of various spin-probes at head and breast after i.v. injection.[21] The decay constants for TEMPOs were several times larger than those for the PROXYLs, and the probes having large n-octanol/water partition coefficients were reduced more quickly. The decay constants at head were significantly smaller than those at breast. The probes shown in Table 2 cannot pass through the blood-brain barrier. The difference in the decay constants between head and breast implies that the nitroxyl reduction occurs not only during blood circulation but also through interaction with tissues.

There should be several different mechanisms in *in vivo* spin-clearance, involving direct reduction to the corresponding hydroxylamine in blood circulation or tissues, decomposition to the other non-paramagnetic compounds, excretion through kidney or liver to urine or feces, the binding

Table 2. *In vivo* decay constants of ESR signals at head and breast of mice after i.v. injection of nitroxyl probes. The decay constants (/min) are presented as mean±S.D. The numbers of experiments are indicated in parentheses.[21]

	k (/min)		n-Octanol/water partition coefficient
	Head	Breast	
TEMPO derivatives			
Hydroxy-TEMPO	0.55±0.13 (5)	0.97±0.33 (5)	4.22
Amino-TEMPO	0.92±0.14 (5)	1.53±0.54 (8)	0.04
CAT-1	0.07±0.05 (5)	0.21±0.08 (6)	0.0004
TEMPO-T	0.12±0.04 (6)	0.27±0.10 (6)	
Carboxy-TEMPO	0.25±0.02 (5)	0.46±0.10 (5)	0.03
Hexamethonium-TEMPO	0.02±0.01 (2)	0.06±0.04 (4)	
PROXYL derivatives			
Carbamoyl-PROXYL	0.05±0.01 (5)	0.10±0.05 (6)	0.68
Carboxy-PROXYL	0.04±0.02 (5)	0.04±0.01 (5)	0.01
Barbiturate-PROXYL	0.04±0.02 (2)	0.06±0.03 (4)	
Sulfonamide-PROXYL	0.12±0.02 (2)	0.20±0.08 (5)	
OXANO derivatives			
2-DOXYL-butane			
(aqueous phase)	0.34±0.07 (6)	0.35±0.10 (5)	
(lipid phase)	0.16±0.07 (6)	0.15±0.03 (5)	

Table 3. Effect of sulfhydryl-blockers having different membrane permeability on reduction of nitroxyl radical in mouse lung. Hydroxy-TEMPO solution (5 mM) containing sulfhydryl-blocker (5 mM) was injected into mouse lung, and first-order reduction rate constants (k) were measured.[18]

Sulfhydryl-blockers	Distance[a] (nm)	k[b] (min^{-1})
None		0.111±0.013 (7)
Membrane permeable		
N-Ethylmaleimide		0.037±0.001 (3)
Floating in membrane		
ε-Maleimidocaproic acid	1.08	0.040±0.002 (3)
γ-Maleimidobutyric acid	0.83	0.076±0.015 (4)
β-Maleimidopropionic acid	0.71	0.120±0.011 (4)
Membrane-impermeable		
p-Chloromercuriphenylsulfonic acid		0.015±0.023 (4)

[a]The distance is that from the reactive maleimide moiety to the carbon atom of the carboxyl group.
[b]Each value represents the mean±S.D. The numbers in parentheses are numbers of animals.

to macromolecules, etc. Excretion of spin-probes through kidney may be a major mechanism, since the accumulation of ESR signal of spin-probes was observed in the bladder and urine.[22]

Signal decay in the early stages may, however, arise from the reduction of nitroxyl radicals to the corresponding hydroxylamine. A small amount of blood was collected from the mouse which was pre-loaded with spin-probes, and the signal intensity in the collected blood was compared before and after oxidation with ferricyanide. The result was the stoichometric restoration of signal intensity to the amount injected.[16] The presence of the corresponding hydroxylamine was also confirmed by thin-layer chromatography of the metabolic products in mouse lung.[18]

There are many reductants such as ascorbic acid towards nitroxyl radicals in blood. The direct contribution of blood components to *in vivo* nitroxyl reduction must be very small, since the decay constants of spin-probes in the collected blood were about 0.1 times those observed in *in vivo* measurements.[6] Interaction of spin-probes with blood vessels, endothelial cells or parenchymal cells of tissues may be important in the *in vivo* reduction of nitroxyl radicals.

Mouse lung also has a reduction system against nitroxyl radical which is unwashable and inactivated by homogenization.[17,18] Nitroxyl radicals without any charges were reduced much more quickly than those having charges. The reduction was inhibited strongly with membrane-permeable or amphiphilic SH-blockers but not with membrane-impermeable blockers (Table 3). The membrane-permeable reducing SH reagent dithiothreitol enhanced the reduction of nitroxyl radicals, although it did not reduce the radicals *in vitro*. These results indicate that the reduction system in mouse lung is located within the hydrophobic region of plasma membrane and that lipophilic SH-modifying compounds regulate the reduction.

Table 4. Decay constants for hydroxy-TEMPO and carbamoyl-PROXYL in whole mice under hypo- and hyperoxia (/min).[13)]

O_2	Hydroxy-TEMPO Abdomen	Hydroxy-TEMPO Head	Carbamoyl-PROXYL Abdomen	Carbamoyl-PROXYL Head
12%	0.95±0.03 ⌉	0.69±0.01	0.12±0.01 ⌉	0.10±0.01 ⌉
20%	0.85±0.01 ⌋ *	0.71±0.01	0.10±0.02 ⌋⌉ **	0.07±0.01 ⌋ ***
80%	0.84±0.02	0.71±0.01	0.15±0.02 ⌋ ***	0.07±0.01

Clearance constants are presented as mean±S.E. over 6 experiments. $^*p<0.1$, $^{**}p<0.05$, $^{***}p<0.001$.

Relation of Nitroxyl Reduction to Biological Phenomena

Table 4 demonstrates decay constants at head and abdomen of mice that were exposed to different oxygen concentrations.[13)] The decay constants of both hydroxy-TEMPO and carbamoyl-PROXYL under 12% oxygen were significantly larger than those under 20% oxygen in both domains. We previously reported that nitroxyl radical loses its paramagnetism more rapidly by the treatment of microsomes under hypoxic condition.[8, 10)] A hypoxic condition may also favor reduction of nitroxyl radicals in living mice.

Decay constant of carbamoyl-PROXYL in the abdomen under 80% oxygen was significantly greater than that under 20% oxygen ($p < 0.001$). Active oxygen species such as $\cdot O_2^-$, $\cdot OH$, and H_2O_2 are reported to be generated in the liver under hyperoxia,[23)] and the nitroxide radical loses its paramagnetism by interaction with active oxygen species.[24)] In fact, the pre-load of antioxidants such as Trolox, uric acid, and glutathione retarded the enhancement of signal decay under hyperoxia (Table 5), and the retardation corresponded to those estimated with TBA-reactive substances. I.p. administration of ascorbic acid at a dose of 10 mg/kg body weight did not show any effect *in vivo*, although it can reduce nitroxyl radical quickly *in vitro*.

Ischemia-reperfusion also enhanced the *in vivo* signal decay.[12, 15)] The reduction of nitroxyl

Table 5. Effect of various antioxidants on decay constants for carbamoyl-PROXYL in abdomen under normoxia and hyperoxia.[14)]

	Dose (mg/kg)	20% Oxygen	(/min) 80% Oxygen
Vehicle 1[1]		0.102±0.012	0.138±0.014
Vehicle 2[2]		0.105±0.015	0.132±0.012
Trolox	1	0.106±0.010	0.114±0.015[4]
Uric acid	10	0.109±0.011	0.104±0.021[5]
Glutathione	10	0.115±0.009[3]	0.109±0.014[6]
Ascorbic acid	10	0.105±0.016	0.135±0.008

Kinetic constants are presented as mean±S.D. over 5 or 6 experiments.
[1]Saline (0.2ml) administered.
[2]Saline containing ethanol (1.4% v/v, 0.2ml) administered.
[3]$p<0.1$, different from vehicle 1. [4]$p<0.1$, different from vehicle 2
[5]$p<0.01$, different from vehivle 1. [6]$p<0.005$, different from vehicle 1.

probes in femoral muscle was significantly larger in the group treated with ischemia-reperfusion than that without ischemia-reperfusion ($p < 0.05$), and the increment of the ratio was inhibited by treatment with SOD or allopurinol, suggesting that the generation of $\cdot O_2^-$ contributes to signal reduction of nitroxyl radicals by ischemia-reperfusion injury.

Aging and feeding also influenced the signal reduction of nitroxyl radicals at head of mice after i.p. administration.[16] The nitroxyl reduction of carbamoyl-PROXYL for young 6-month-old mice fed *ad libitum* was significantly faster than that for 30-month-old mice fed *ad libitum*, and the reduction rate for 39-month-old, food-restricted mice was close to that for 6-month-old mice fed *ad libitum*. These results suggest that reducing capacity declines in older mice and that food restriction prevents its age-dependent retardation.

Conclusions

In this chapter, we demonstrate that an *in vivo* ESR-CT system with nitroxyl radicals as a probe makes it possible to estimate *in vivo* free radical reactions in living animal non-invasively, since the nitroxyl radicals are susceptible both to biological redox state and to active oxygen species. It has been found that the *in vivo* signal decay of nitroxyl radicals is influenced by physiological and pathological phenomena such as aging, inspired oxygen concentration, ischemia-reperfusion injury, etc. The present paper strongly suggests that *in vivo* ESR measurement with nitroxyl radical as a probe is a very useful technique to estimate the influence of antioxidants on biological radical reaction in the living body.

Drug delivery system (DDS) is being actively investigated using radio-active compounds in the field of pharmaceutical sciences. *In vivo* ESR technique should prove to be a powerful method for DDS investigations, since it is non-invasive and occurs in real time. Recently, we succeeded using *in vivo* ESR technique to reveal not only pharmacokinetics but also metabolism of the lipid microsphere having spin-labeled triglyceride in mice after i.v. injection.[25] *In vivo* ESR-CT technique has great advantages in the investigation of the stability and the pharmacokinetics of DDS in living body.

Acknowedgment

This work was in part supported by Grant-in-Aids for Research on Priority Area, for Co-operative Research, for General Scientific Research, and for Developmental Scientific Research from the Ministry of Education, Science, Sports and Culture of Japan, by Special Coordination Funds of the Science and Technology Agency of the Japanese Government, by the Naito Foundation, and by the Cosmetology Research Foundation.

References

1. Nishikawa, H., Fujii, H., and Berliner, L. J.; *J. Mag. Res.* (1985) 62, 79-86
2. Subczynski, W. K., Lukiewica, S., and Hyde, J. S.; *Magn. Reson. Med.* (1986) 3, 747-754
3. Bacic, G., Nilges, M. J., Magin, R. L., Walczak, T., and Swartz, H. M. *Magn. Reson. Med.*

(1989) 10, 266-272

4. Ishida, S., Kumashiro,H., Tsuchihashi,N., Ogata,T., Ono,M., Kamada,H., and Yoshida, E.; *Phys. Med. Biol.* (1989) 34, 1317-1323
5. Ferrari, M., Colacicchi, S., Gualtieri, G., Santini, M. T., and Sotgiu, A.; *Biochem. Biophys. Res. Commun.* (1990) 166, 168-173.
6. Utsumi, H., Muto, E., Masuda, S., and Hamada, A.; *Biochem. Biophys. Res. Commun.* (1990) 172, 1342-1348.
7. Eaton, G.R., Eaton, S.S., and Ohno, K.; *EPR Imaging and In Vivo EPR,* CRC Press, New York, 1991
8. Miura, Y., Utsumi, H., and Hamada, A.; *J. Biochem.* (1990)108, 516-518.
9. Miura, Y., Utsumi, H., and Hamada A.; *Arch. Biochem. Biophys.* (1993) 300, 148-156
10. Utsumi, H., Shimakura, A., Kashiwagi, M., and Hamada, A.; *J. Biochem.*. (1989) 105, 239-244
11. Utsumi, H., Masuda, S., Muto, E., and Hamada, A.; In: *Oxidative Damage & Repairs.* K. J. A. Davies (Ed), Pergamon Press, New York, (1991) pp.165-170
12. Utsumi, H., Takeshita, K., Miura, Y., Masuda, S., and Hamada, A.; *Free Rad. Res. Commun.*, (1993)19, s219-225
13. Miura, Y., Utsumi, H., and Hamada, A.; *Biochem. Biophys. Res. Commun.* (1992) 182, 1108-1114
14. Miura, Y., Hamada, A., and Utsumi, H.; *Free Rad. Res. Commun.* (1995) 22, 209-214
15. Masuda, S., Utsumi, H., and Hamada, A. In: *Oxygen Radicals.* K. Yagi, M. Kondo, E. Niki, T. Yoshikawa (Eds.), Excerpta Medica, Amsterdam, (1992) pp.175-178
16. Gomi, F., Utsumi, H., Hamada, A., and Matsuo, M.; *Life Science* (1993) 52, 2027-2033
17. Takeshita, K., Utsumi, H., and Hamada,A.; *Biochem. Biophys. Res. Commun.* (1991) 177, 874-880
18. Takeshita, K., Utsumi, H., and Hamada A.; *Biochem. Mol. Bio. Int.* (1993) 29, 17-24
19. Masumizu, T., Tatebe, T., Masuda, S., Muto, E., Utsumi, H., and Hamada, A.; *Jpn. Mag. Reson. Med.* (1991) 2, 63-68
20. Kocherginsky, N. and Swarzt, H. M.; *Nitroxide Spin Labels,* CRC Press, New York, 1995
21. Sano, H., Chignell, C. F., Utsumi, H.; submitted
22. Masuda, S., Utsumi, H., and Hamada, A.; *Jpn. Mag. Reson. Med.* (1991) 12, 69-74
23. Nishiki, K., Jamieson, N., Oshino, N., and Chance, B.; *Biochem. J.* (1976) 160, 343-355.
24. Samuni, A., Krishna, C. M., Riesz, P., Finkelstein, E., and Russo, A.; *J. Biol. Chem.* (1988) 263, 17921-17924.
25. Yamaguchi, T., Itai, S., Hayashi, H., Soda, S., Hamada, A., and Utsumi, H.;*Pharm. Res.* (1996) 13, 729-733

3. Biodefence against Active Oxygens and Free Radicals Induced by Oxidative Stress

Mikinori KUWABARA and Osamu INANAMI

Laboratory of Radiation Biology, Graduate School of Veterinary Medicine, Hokkaido University, Sapporo 060, Japan

A chemical antioxidant, α-phenyl-N-*tert*-butylnitrone (PBN), and natural antioxidants, extracts from Rooibos tea and green tea, were employed as biodefense agents against oxygen radical-derived oxidative stress, that is, ischemia/reperfusion-induced injury in the rat brain, copper-induced hepatitis in the rat liver and morphological changes as well as lipid peroxidation in the rat brain with aging. Four repetitive i. p. injections of PBN at a relatively low concentration (30 mg/kg) to gerbils before and after ischemia were found to markedly suppress neuronal death in the hippocampal CA1 region after 5-min ischemia and subsequent reperfusion. The chronic subcutaneous administration of 128 mg/kg of PBN every two days to Long-Evans Cinnamon rats, which are autosomal recessive mutants that accumulate an abnormal amount of copper ions in the liver, effectively prolonged the initiation of fulminant hepatitis with jaundice. The protective effects of oral *ad libitum* administration of extracts from *Aspalathus linearis* (Rooibos tea) to Wistar rats from 3 to 24 months against damage to the central nervous system accompanying aging were proved by measuring age-related accumulation of lipid peroxides and observing morphological changes in the brain. When (-)-catechin, which is an extract from green tea, was orally administered to gerbils *ad libitum* at a concentration of 1.0 mg/mL for 2 weeks before ischemia and for 1 week after ischemia, its ability to protect against ischemia/reperfusion-induced neuronal death in the hippocampal CA region was observed.

Introduction

Transient ischemia in the brain results in the selective loss of hippocampal CA1 pyramidal cells in Mongolian gerbils [1] and rats [2]. Several reports have demonstrated that the neurotoxic reactive oxygen intermediates (ROIs) O_2^-, H_2O_2, OH and NO generated during cerebral ischemia/reperfusion are directly associated with this injury and that ischemia-induced activation of the hypoxanthine/xanthine oxidase system is mainly responsible for ROI formation within 1 hr after reperfusion [3-6]. On the other hand, accumulation of microglia in the CA1 region is observed 2 days after ischemia/reperfusion [7]. This suggests the possibility that microglia cells release ROIs for several days. Therefore, it is important to examine the implication of ROIs produced not only within 1 hr but also over a certain period of time in delayed neuronal death. For this purpose, PBN was administered by i. p. injection once 1 hr before ischemia and three times after ischemia, and its protective effect against neuronal death in the CA1 region was histologically evaluated because it is known to serve as a chemical antioxidant [8,9]. Characteristics of PBN as an antioxidant were further examined by observing its protective effects against copper-induced rat fulminant hepatitis with jaundice [10]. The Long-Evans Cinnamon (LEC) rat is a mutant strain with hereditary hepatitis [11]. An excess amount of copper is accumulated in the liver of this rat [12]. Free or metallothionein-bound copper ions produce ROIs which are thought to be responsible for the acute-phase hepatitis [13]. PBN administration to LEC rats was, therefore, expected to demonstrate protective effects against liver injury.

The protective effects of natural antioxidants orally administered to rats against ROI-derived oxidative stress were examined in age-related deterioration of the central nervous system (CNS) and ischemia/reperfusion-induced injury in the rat brain. In aging processes, it is believed that the age-related deterioration of the CNS is partly due to the cytotoxicity of ROIs generated in the brain [14,15]. The abnormal increase in lipid peroxides accompanying aging is assumed to cause irreversible damage to the CNS. Since Rooibos tea made of leaves of *Aspalathus linearis* was found to have properties to scavenge ROIs and to protect against the formation of thiobarbituric acid reactive substances (TBARS), its oral administration to rats for a long period of time was tested to determine whether this natural substance protected against the deterioration of the CNS accompanying aging. Similarly, oral administration of (-)-catechin, which is a flavonoid extracted from free tea, to gerbils for 21 days before and after ischemia was carried out to examine its protective effects against ischemia/reperfusion-induced neuronal death.

Materials and Methods

Attenuation of cerebral ischemia injury by PBN in gerbils

Mongolian gerbils (*Meriones unguiculatus*) weighing 65-90g were used. They were provided with food and water *ad libitum*. PBN (Aldrich Chemical Co.) was dissolved in sterile saline solution and administered to rats by the following protocols: 1) a single i. p. administration of PBN at a dose of 200 mg/kg 1 hr before ischemia, 2) repetitive i. p. administrations of PBN at a dose of 30 mg/kg 1 hr before and 20, 44, 68 hrs after ischemia, 3) preadministration of saline without PBN, and 4) a sham operation. For ischemia operation, bilateral common carotid arteries were occluded for 5 min under pentobarbital anesthesia. The sham operation was performed using the same procedure without occlusion. Seven days after ischemia, coronal sections 12 mm thick including the dorsal hippocampal region were made and stained with cresyl violet and examined under a light microscope. The cellular damage in the CA1 region was scored by counting neurons per 1 mm of linear length in the stratum pyramidale of the CA1 region.

Prolongation of the initiation of copper-induced fulminant hepatitis by PBN administration in LEC rats

LEC rats maintained at the Institute for Animal Experimentation, Hokkaido University School of Medicine, were used for experiments. Every other day PBN dissolved in sesame oil at a concentration of 128 mg/kg was subcutaneously administered to female rats, beginning at the age of 13 weeks and continuing for 30 weeks. Evaluation of lipid peroxidation in the liver was carried out by measuring the thiobarbituric acid-reactive substances (TBARS) according to the method of Uchiyama and Mihara [16]. Tissue blocks from each animal were fixed with 4% paraformaldehyde solution in 0.1 M phosphate buffer (pH 7.3) overnight at 4°C. Paraffin sections (3 mm) were stained with hematoxylin-eosin.

Suppression of age-related accumulation of lipid peroxides in rat brain by orally administered Rooibos tea

Fermented and dried leaves of *Aspalathus linearis*, which is a leguminous plant from South Africa, were boiled in water (2 g/L) for 20 min. The solution is called Rooibos tea, which mainly contains flavonoid glycosides falvone and flavonol, minerals and ascorbic acid. Three groups of female Wistar rats (WKAH) were prepared: (1) 5-week-old rats, (2) 24-month-old rats fed with water and (3) 24-month-old rats fed with Rooibos tea from 3 months after birth. The effects of oral administration of Rooibos tea on the accumulation of lipid peroxides, as well as the morphological changes in the brain caused by aging, were examined by measuring the amount of TBARS and by examining the brain with a magnetic resonance imaging (MRI) technique [17, 18]. T_1-weighted proton MR images were obtained on an Oxford 7.05 T magnet equipped with a Toshiba workstation with an SIS300/183 imaging system using a two-dimensional Fourier transform technique.

Attenuation of cerebral ischemia injury by orally administered (-)-catechin in gerbils

Mongolian gerbils weighing 65-90 g were used. They were provided with food *ad libitum*. (-)-Catechin was dissolved in sterile water at a concentration of 1.0 mg/mL and orally administered *ad libitum* to rats instead of water for 2 weeks prior to ischemia and 1 week after ischemia. Surgery for transient focal brain ischemia and histological examinations were performed as above. Superoxide scavenging activity of the brain was assessed by a method combining ESR and spin trapping with 5,5-dimethyl-1-pyrroline *N*-oxide (DMPO). The brain homogenate was mixed with DMPO in the superoxide generating system consisting of hypoxanthine/xanthine oxidase. The adducts between superoxide and DMPO were quantitatively measured by ESR. The suppressive effect of the brain homogenate on the generation of the adducts was estimated as the superoxide-scavenging ability .

Results and Discussion

Attenuation of cerebral ischemia injury by PBN in gerbils

Photographs of coronal sections containing the hippocampal CA1 region 7 days after occlusion of the bilateral carotid artery for 5 min in the treated and sham-operated animals are shown in Fig. 1. Compared to the sham-operated group, a considerable number of pyramidal cells in the CA1 region were found to disappear after ischemia/reperfusion in the group injected with saline without PBN (Figs. 1A and 1B). Four repetitive injections of 30 mg/kg of PBN to rats at -1 hr before and 20, 44, 68 hrs after the ischemia clearly conserved the morphological shape of the CA1 region containing pyramidal cells (Fig. 1C).

The percentages of surviving cells are summarized in Fig. 2, in which the sham-operated group was scored as 100%. The surviving fraction after ischemia/reperfusion without PBN was 17.0±4.3%. When a massive dose of PBN (200 mg/kg) was administered by a single injection 1 hr before ischemia, a large number of the cells survived (78.4±13.2%), whereas with four repetitive injections of 30 mg/kg of PBN to rats after the ischemia almost 100% of cells survived (96.6±3.0%). From these results it was concluded that not only ROIs produced immediately after ischemia but also the delayed biochemical reactions persisting for a few days after ischemia were related to neuronal death. However, a single preadministration of 30 mg/kg 1 hr before ischemia as well as administrations of 30 mg/kg of PBN at 1 hr before and 20 hrs after ischemia resulted in unsatisfactory effects (Fig. 1D), suggesting that the pretreatment of rats with PBN is essential for complete protection against neuronal cell damage.

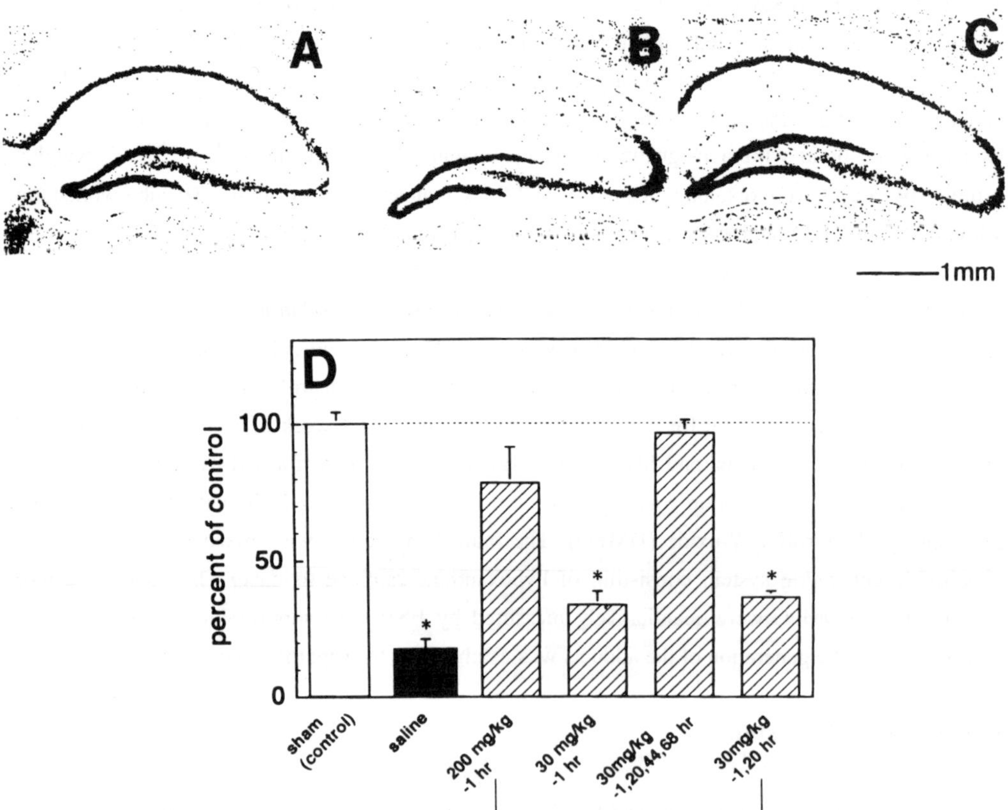

Fig. 1. Photographs of coronal sections in the hippocampal CA1 region 7 days after 5-min ischemia; (A) sham, (B) saline and (C) PBN (30 mg/kg i. p.) 1 hr before and 20, 44 and 68 hrs after ischemia. Magnification x 10. (D) The effects of PBN injection on hippocampal CA1 neuronal death. Columns show percentages of the sham operation. All columns and vertical bars represent the mean ± SD. *: $p<0.05$.

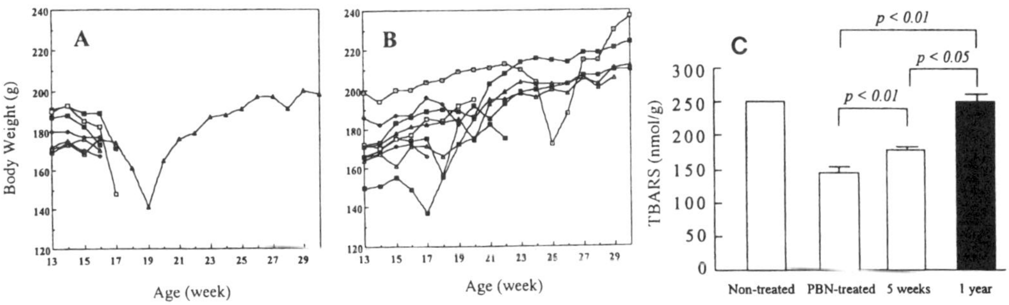

Fig. 2. Age-dependent changes of body weights in LEC rats. Disappearance of lines means that rats died with jaundice. (A) non-PBN-administered, and (B) PBN-administered rats. (C) TBARS of liver homogenates obtained from four groups of rats. Values except for nontreated rats represent mean ± SD.

Accumulation of microglial cells is observed in the hippocampal region 2 days after ischemia/reperfusion and the possibility of ROI release from microglia has been suggested [19,20]. PBN delivered by repetitive administrations after ischemia may play a protective role by scavenging microglia-induced ROIs. However, double injections of PBN at 20 and 44 hrs after ischemia brought about little protection against neuronal death (data not shown). This means that the protection against neuronal death by PBN administration may not necessarily be due to its scavenging ability. Our preliminary experiments showed that proteins having superoxide-scavenging ability were induced in the brain by repetitive administrations of PBN (unpublished results). The chronic administration of a small amount of PBN was more effective and practical than a single preadministration of a massive dose of PBN not only for protecting against ischemia/reperfusion-induced cell injury but also for avoiding the acute side-effects of PBN.

Prolongation of the initiation of copper-induced fulminant hepatitis by PBN administration in LEC rats

The effects of PBN administration to 13- to 30-week-old LEC rats on body weight were measured first (Figs. 2A and 2B). In non-PBN-administered rats, no increase in body weight in eight of nine rats was observed for 16-17 weeks, and they died with 18 weeks. For one rat, the body weight reached a minimum at 19 weeks and gradually increased up to 30 weeks. This rat survived for more than 30 weeks (Fig. 2A). One of the nine PBN-administered rats died at 17 weeks of age with no increase in body weight; two survived for 19 and 20 weeks without loss of body weight; one showed loss of body weight at 17 weeks but survived for 22 weeks and one survived 29 weeks without loss of body weight. A total of four rats survived for more than 30 weeks (Fig. 2B). These data gave the survival times of non-PBN- and PBN-treated rats. The mean survival times of non PBN-administered and PBN-administered rats were 15.3±1.5 weeks and 21.4±4.6 weeks, respectively. The mean survival time of PBN-administered rats was significantly prolonged.

Prolongation of the maximal activities of GOT and GTP in sera of PBN-treated rats was also noted (data not shown).

The LEC rat is a mutant strain with hereditary hepatitis in which an excess amount of copper is accumulated in the liver [11,12]. Since copper is thought to generate ROIs, TBARS of the liver tissues from no-PBN- and PBN-administered rats were measured as a result of oxidative reactions (Fig. 2C). Since only one rat was available as a 30-week survivor in the non-PBN-treated group, the TBARS value was determined to be 250±13 nmol/g from the group of 1-year survivors. The TBARS level of the PBN-treated rats was 145±10 nmol/g, significantly lower than that of the non-PBN-treated rats.

The histological changes in the livers of non-PBN- and PBN-treated rats were finally examined with a light microscope. The liver showed massive and multifocal necrosis with filtration of a few inflammatory cells and giant hepatocytes. Proliferation of peudobile ducts and increases numbers of Kupffer cells phagocytizing were observed. In PBN-administered LEC rats, only minor morphological changes were observed in the liver (Fig. 3).

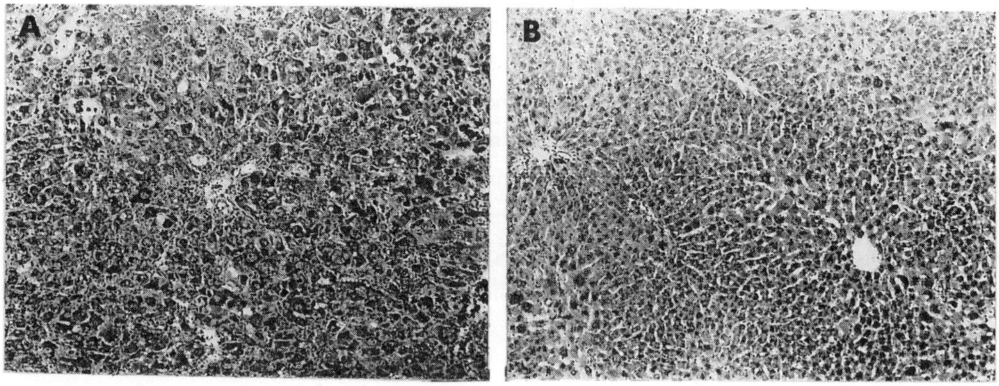

Fig. 3. Photographs of LEC rat livers. (A) 30-week survivor with non-PBN administration, and (B) 30-week survivor with PBN administration. Magnification x 75.

In vivo PBN administration caused a significant decrease in lipid peroxidation in the liver and prevented hepatic cell injury with fulminant hepatitis. This seemed to be explained by its ability to trap ROIs as well as ROI-induced lipid radicals. Thus, PBN was found to have potential therapeutic value for hepatitis.

Suppression of age-related accumulation of lipid peroxides in rat brain by administration of Rooibos tea

The protective effects of a natural antioxidant, Rooibos tea, orally administered to rats, against ROI-derived oxidative stress were examined in the age-related deterioration of the central nervous system (CNS). Fig. 4 summarizes the contents of TBARS in eight different brain regions in 5-week-old Wistar rats, and in 24-month-old rats with and without Rooibos tea. The contents of TBARS in the frontal cortex, occipital cortex, hippocampus and cerebellum of the aged group without Rooibos tea were found to be significantly increased compared to those of the 5-week-old group. By contrast, the contents of TBARS in these regions in the aged group with Rooibos tea were lower than those in the aged group without Rooibos tea, and were quite similar to those in the young group.

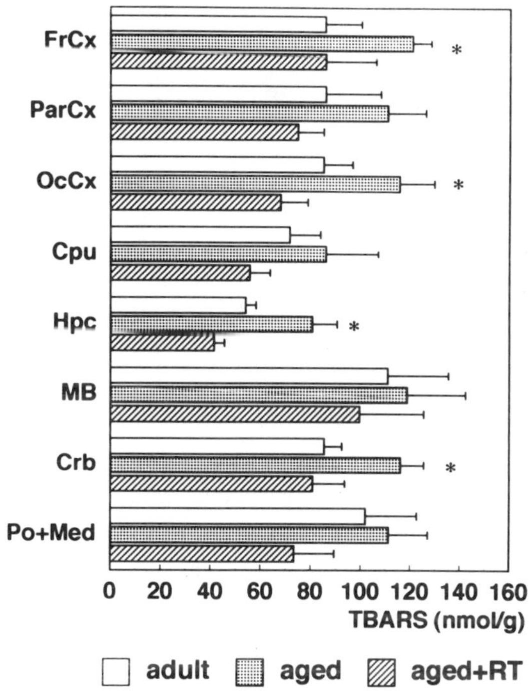

Fig. 4. The effects of oral administration of Rooibos tea on age-related accumulation of lipid peroxides in the eight brain regions. *p<0.05, significant different from TBARS in 5-week-old group.

The effects of the increase in lipid peroxide level on the morphological changes of the brain were examined by employing an MRI apparatus designed for use with small laboratory animals. Fig. 5 shows MR images of coronal sections at +5.5 mm from the interaural line in the brain of a young rat, and in aged rats with and without Rooibos tea. In the MR image of the young rat, the cerebral cortex, hippocampus, thalamus and hypothalamus were clearly recognized (Fig. 5A). Accompanying aging, the MRI signal intensities in these regions significantly decreased (Fig. 5B). Fig. 5C shows an MR image of the brain of a rat administered Rooibos tea. The decrease in MRI intensity in the cerebral cortex, thalamus and hippocampus was small. The results showed a tendency similar to those from TBARS. Therefore, we concluded that a decrease in MRI signal intensity reflected lipid peroxidation. Since Rooibos tea contains several flavonoids, these are thought to function as free radical scavengers in the brain and suppress lipid peroxides.

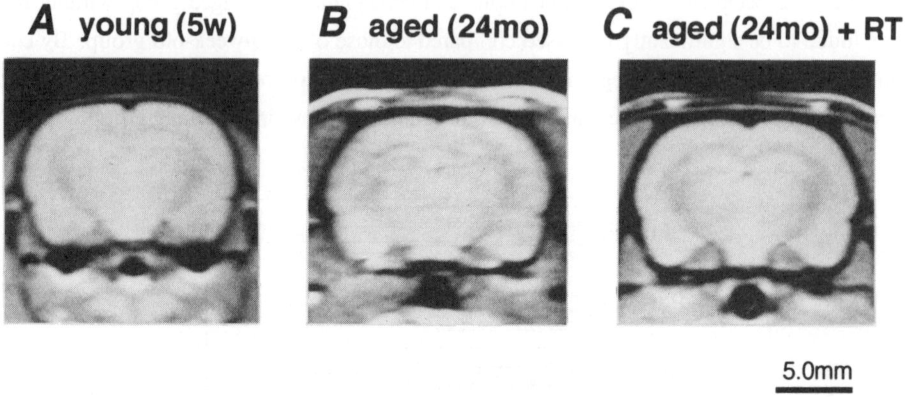

Fig. 5. T_1-weighted MR images of coronal sections of rat heads (1-mm thick) at +5.5 mm from the interaural line. (A) 5-week-old rat, (B) 24-month-old rats and (C) Rooibos tea-administered rats (24 months old). A magnetic field 7.05 T, T_R = 500 ms, T_E = 20 ms, FOV = 6 x 6 cm^2, pixel size = 0.12 x 0.12 mm^2, 128 phase encodes, 8 acquisitions.

Attenuation of cerebral ischemia injury in gerbils by orally administered (-)-catechin

As a second natural antioxidant, (-)-catechin purified from green tea was employed and the effects of oral administration on neuronal death induced by ischemia/reperfusion in gerbils examined. Administration of (-) -catechin begun at 14 days before ischemia and continued until 7 days after ischemia was roughly estimated as resulting in an uptake of 3.5 mg/day. After coronal sections containing the hippocampal CA1 region of the brain were made and stained, neurons along the stratum pyramidale were counted (Fig. 6A). The results showed that ischemia/reperfusion-induced neuronal death was clearly inhibited by oral-administration of (-)-catechin. To clarify its effects on brain injury, the superoxide-scavenging activities of brain homogenates were examined by the spin-trapping method with DMPO (Fig. 6B). Brain homogenates from rats without (-)-catechin showed superoxide-scavenging ability to a certain extent, whereas brain homogenates from rats with catechin scavenged more superoxides. This led us to conclude that oral administration of (-)-catechin increased the superoxide-scavenging ability of the rat brain. The increase is thought to be responsible for the protection against ischemia-induced neuronal death. Catechin itself has superoxide-scavenging ability, but the results also suggest that substances having superoxide-scavenging ability are induced in the brain by oral administration .

Conclusions

From the experiments presented here, it was concluded that the protective abilities of chemical and natural agents against oxidative stress are mainly explained by their antioxidative properties, but since antioxidative substances are known to sometimes interfere with cellular redox regulation and to alter expression of some genes, resulting in the induction or suppression of protein synthesis [21]), the biodefense effects of these agents on cellular damage originating from oxidative stress should be analyzed on the basis of the concept of signal transduction mechanisms in future studies.

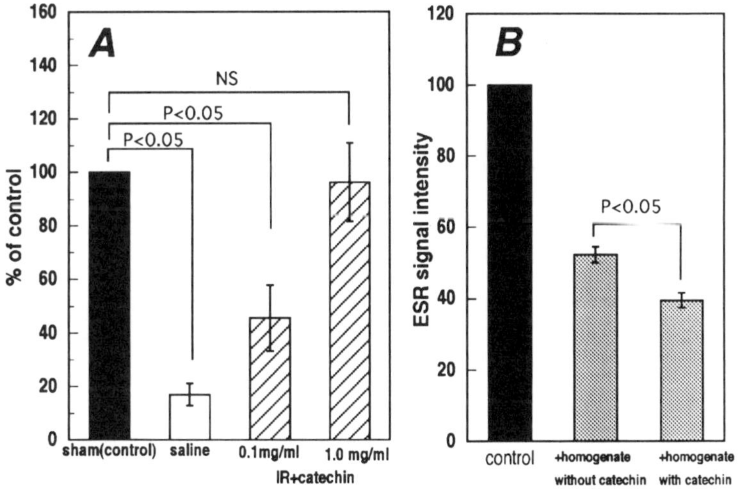

Fig. 6. (A) The effects of (-)-catechin on hippocampal CA1 neuronal death. Columns show percentages of the sham operation. All columns and vertical bars represent the mean ±SD. *: $p<0.05$. (B) Superoxide-scavenging activities of brain homogenates in catechin-administered rats. The amount of superoxides generated by the xanthine/hypoxanthine system was regarded as 100%.

References

1. Kirino, T., *Brain Res.*, 1982, **239**: 57-69.
2. Pulsinelli, W. A., Brierley, J. B. and Plum, F., *Ann. Neurol.*,1982, **11**: 491-498.
3. Sakamoto, A., Ohnishi, S. T., Ohnishi, T. and Ogawa, R., *Brain Res.*,1991, **554**: 186-192.
4. Hysilop, P. A., Zhang, Z., Pearson, D. V. and Prebus, L. A., *Brain Res.*,1995, **671**: 181-186.
5. Phillis, J. W., Sen, S. and Cao, X., *Neurosci. Lett.*,1994, **169**: 88-190.
6. McCord, J. M., *N. Engl. J. Med.*,1985, **312**: 159-163.
7. Hu, S., Sheng, W. S., Peterson, P. K. and Chao, C. C., *Glia*,1995, **13**: 45-50.
8. Phillis, J. W. and Clough-Helfman, C., *Neurosci. Lett.*,1990, **116**: 315-319.
9. Clough-Helfman, C. and Phillis, J. W., *Free Rad. Res. Comm.* 1991, **15**: 177-188.
10. Yamashita, T. Ohshima, H., Asanuma, T., Inukai, N., Miyoshi, I., Kasai, N., Kon, Y., Watanabe, T., Sato, F., and Kuwabara, *M. Free Rad. Biol. Med*, 1996, in press.
11. Yoshida, M. C., Masuda, R., Sasaki, M., Takeuchi, N., Kobayashi, H., Dempo, K., and Mori, M., *J. Hered*, 1987, **78**: 361-365.
12. Masuda, R., Yoshida, M. C., Sasaki, M., Dempo, K., and Mori, M., *Jpn. J. Cancer Res.* 1988, **79**: 823-835.
13. Hiyamuta, S. and Takeuchi, N., *Biochim. Biophys. Res. Commun.*, 1993, **197**: 1140-1145.
14. Harman, D., *Proc. Natl. Acd. Sci. USA*, 1981, **78**: 7124-7128.
15. Floyd, R. A., *Perspectives*, 1991, **13**: 1597.
16. Uchiyama, M. and Mihara, M., *Anl. Chem.*, 1978, **86**: 271-278.

17. Kuwabara, M., Asanuma, T., Inanami, O., Jin, T., Shimokawa, S., Kasai, N., Kator, K., and Sato, F., *J. Vet. Med. Sci.*, 1994, **55**: 933-938.
18. Inanami, O., Asanuma, T., Inukai, N., Jin, T., Shimokawa, S., Kasai, N., Nakano, M., Sato, F., and Kuwabara, M., *Neurosci. Lett.*, 1995, **196**: 85-88.
19. Kato, H., Kigure, K., Araki, T., and Itoyama, Y., *Brain Res.*, 1994, **664**: 69-76.
20. Hu, S., Sheng, W. S., Peterson, P. K., and Chao, C. C., *Glia*, 1995: **13**: 45-50.
21. Burdon, R. Y., *Free Rad. Biol. Med.*, 1995, **18**: 775-794.

4. α-Lipoic Acid: A Metabolic Antioxidant and Redox Modulator of Signal Transduction

Lester Packer

Department of Molecular and Cell Biology, University of California, Berkeley, CA, USA

I. Introduction

α–Lipoic acid (1,2 dithiolane 3 pentanoic or thioctic acid) is a potent thiol antioxidant. Lipoate is an essential cofactor bound to the E2 enzyme of α–keto acid dehydrogenase complexes, and in the glycine cleavage system of mitochondria. Thus it serves an essential role in carbohydrate, protein and energy metabolism. Lipoate given exogenously is readily absorbed, transported into cells and tissues, and reduced by NADH and/or NADPH dependent enzymes to dihydrolipoate (6,8 dimercapto octanoic acid). Lipoate and dihydrolipoate are medium-chain fatty acid-like substances whose redox potential is lower than glutathione, and whose molecular structure allows for their uptake by all tissues thus far investigated, including the brain, and distribution into all cell compartments. Thus, it has turned out to be an effective tool to investigate antioxidant redox regulation involved in oxidatively regulated pathways for signal transduction and in the treatment of oxidatively induced injury.

II. Antioxidant Actions

A. Lipoate

α–Lipoic acid scavenges hydroxyl radicals, hypochlorous acid, singlet oxygen, and nitric oxide. In addition, it chelates a number of transition metals.

α–Lipoic acid is a potent hydroxyl radical scavenger. α–Lipoic acid (1 mM) completely eliminated the Fenton (2mM H_2O_2 + 0.2 mM $FeSO_4$) derived DMPO-OH (DMPO, 5,5-dimethylpyrroline-N-oxide) adduct signal (1). Another study (2), using a similar hydroxyl radical-generating system (2.8 mM H_2O_2, 0.05 mM $FeCl_3$, 0.1 mM EDTA, and 0.1 mM ascorbate) but a different assay for the radical (deoxyribose degradation) also found α–lipoic acid to be hydroxyl radical scavenger. In this study, a rate constant of $4.7 \times 10^{10}\ M^{-1}s^{-1}$ for the scavenging reaction was determined.

Recent studies from our laboratory with a novel hydroxyl radical producing compound, N,N'-bis(2-hydroperoxy-2-methoxyethyl)-1,4,5,8-naphthalene-tetracarboxylic diimide (NP-III), confirm that α–lipoate is a hydroxyl radical scavenger, and eliminate the possibility that the effect is simply due to metal chelation (3). NP-III produces hydroxyl radicals upon illumination with UVA light. The production of hydroxyl radical was followed by measuring oxidation of bovine serum albumin (BSA) and apolipoprotein B (apoB) of human low-density lipoprotein, as well as by using ESR to detect DMPO-OH spin adducts. At a concentration of 1-2 mM, α–lipoate completely eliminated hydroxyl radical production as assessed by DMPO-adducts, and protein carbonyl formation in apoB and BSA; the rate constant for reaction with hydroxyl radical was found to be $1.92 \times 10^{10}\ M^{-1}s^{-1}$, comparable with that found by Scott et al. (2), essentially a diffusion-limited rate.

There is similar agreement about the ability of α–lipoic acid to scavenge hypochlorous acid. Haenen and Bast (4) and Scott et al. (2) independently found that 50 μM α–lipoic acid almost completely abolished the inactivation of α_1-antiproteinase by 50 μM HOCl. This is in contrast to the glutathione redox couple. Although reduced glutathione is a potent scavenger of HOCl, the oxidized form glutathione disulfide is not (4). The authors of this study speculate that the greater reactivity of α-lipoic acid compared to glutathione disulfide may be due to the somewhat strained conformation of the 5-membered ring in the intramolecular disulfide form of α-lipoic acid. There is no such strain on the intermolecular disulfide of glutathione disulfide perhaps explaining its lack of reactivity in this system.

α–Lipoic acid has been reported to scavenge singlet oxygen in at least four different systems. Two early studies showed that α–lipoic acid reacted with singlet oxygen generated by rubrene autooxidation (5) or by photosensitized oxidation of methylene (6). These experiments were carried out in organic solvents. Later studies conducted under more physiological conditions also indicate that α–lipoic acid is an effective scavenger of singlet oxygen. Kaiser et al. (7) generated singlet oxygen by thermolysis of an endoperoxide and detected it by chemiluminescence. In this system α–lipoic acid reacted with singlet oxygen with a rate constant of $1.38 \times 10^{8}\ M^{-1}s^{-1}$. Also, in experiments in which singlet oxygen was generated by thermolysis of endoperoxide and detected by single strand DNA breaks, α–lipoic acid was confirmed to be a scavenger of singlet oxygen (8).

In addition, it has recently been observed that α–lipoate decreases nitric oxide production in macrophages (H. Kobuchi, personal communication). RAW 264.7 macrophages were co-incubated with lipopolysaccharide and various doses of α–lipoate for 24 hours. Nitrite in the culture medium was measured using the Griess reagent. At doses of α–lipoate above 200 μM there was a significant effect on NO production, with over 50% inhibition at 0.5 mM. It is not possible in this system to determine whether the inhibition was due to direct interactions with NO, to decreased induction of inducible nitric oxide synthase, or to some other effect of α–lipoate. However, in experiments in which NO was produced directly by sodium nitroprusside, α–lipoate decreased NO generation.

α–Lipoic acid may also exert an antioxidant effect in biological systems through transition metal chelation. It has been found to decrease Cd^{2+}-induced toxicity in isolated hepatocytes, although the authors speculate that the effect was due to the conversion of α−lipoic acid to DHLA, which was the true chelating agent (9). Two studies indicate that α−lipoic acid may chelate iron, while one indicates that it does not. Devasagayam et al. (8) observed that α−lipoic acid was effective in quenching singlet oxygen-induced DNA single strand breaks. They also reported that the efficacy of α-lipoate was lower when EDTA was added to the system, indicating that part of the effect was due to iron chelation. Scott, et al. (2) found that α−lipoic acid inhibited the site-specific degradation of deoxyribose by a $FeCl_3/H_2O_2$/ascorbate system, indicating that it was able to remove iron ion bound to deoxyribose. In contrast, in a rat liver microsomal system in which lipid peroxidation was induced by Fe_3SO_4 (10), α−lipoic acid did not inhibit peroxidation. α-Lipoate forms stable complexes with Mn^{2+}, Cu^{2+}, Zn^{2+}, with the complex being almost entirely with the carboxylate group (11). These investigators also examined the complexes formed with bisnorlipoate and tetranorlipoate, two and four carbon shorter homologues of α-lipoic acid, respectively. These were found to be more stable than those with α-lipoic acid, presumably because the shorter chain allows the dithiolane ring to also participate in chelation. The effects of α-lipoic acid on copper have recently been extended to oxidation situations. α-Lipoic acid was effective in preventing Cu^{2+}-catalyzed ascorbic acid oxidation, increased the partitioning of Cu^{2+} into n-octanol from aqueous solution, and inhibited Cu^{2+}-catalyzed liposomal peroxidation (12). These observations indicate that LA is a copper chelator.

B. Dihydrolipoate

With a redox potential of -0.32 V (13) for the DHLA/LA redox couple, dihydrolipoic acid is a potent reductant; for comparison, the redox potential of the GSH/GSSG couple is -0.24 V (14). DHLA reduces GSSG to GSH, but GSH is incapable of reducing α−lipoic acid to DHLA (15). DHLA scavenges hypochlorous acid, and peroxyl, superoxide, hydroxyl, and nitric oxide radicals.

In the systems in which α−lipoic acid was tested for reaction with hypochlorous ion, DHLA was also found to be effective, with about the same scavenging ability (4). DHLA is an effective peroxyl radical scavenger in several different systems. Kagan et al. (16) used AAPH or AMVN to generate peroxyl radicals in aqueous or lipid systems and found that DHLA scavenged peroxyl radicals in these systems. It was also found to scavenge peroxyl radicals (Suzuki et al. 1993) and $CCl_3O_2\cdot$ radicals generated in aqueous solution (Scott et al. 1994). In the latter system a rate constant of 2.7 x 10^7 $M^{-1}s^{-1}$ was calculated.

Using an NP-III system similar to that used with α−lipoate (17), we found that DHLA was also effective to scavenge hydroxyl radical completely eliminating the ESR DMPO-OH signal produced upon illumination of NP-III with UV light at a dose of 500 μM.

In a study in which $O_2^{\bullet-}$ was generated by xanthine-xanthine oxidase and detected by ESR using DMPO spin trap (1), DHLA was found to react with superoxide radical with a rate constant of 3.3 x 10^5 $M^{-1}s^{-1}$. Reaction was confirmed by assaying the decreasing sufhydryl content of DHLA and by detecting an increase in H_2O_2. These results were confirmed by this group in a later study (18) using the same system for generating superoxide radical and competition with epinephrine oxidation to assess DHLA's scavenging ability. In this study, a rate constant of $7.3 \times 10^5\ M^{-1}s^{-1}$ was found in good agreement with the previous work. In contrast to these studies, Scott et al. (1994), using a similar superoxide generating system (hypoxanthine-xanthine oxidase) and detecting superoxide by $O_2^{\bullet -}$-dependent nitro-blue tetrazolium reduction, found no scavenging effect of DHLA. It is difficult to explain the discrepancy in the results of the two groups. However, a recent study in which superoxide was detected by chemiluminescence showed DHLA to be a superoxide scavenger with a rate constant of $1.4 \times 10^5\ M^{-1} \bullet s^{-1}$ (19). Hence, the majority of studies, in several different systems, indicate that DHLA is a superoxide scavenger. Similar to α–lipoate, it has been observed that DHLA reacts with nitric oxide (our unpublished finding).

C. Redox interactions with other antioxidants

DHLA appears to be able to regenerate other antioxidants, such as ascorbate and (indirectly) vitamin E, from their radical forms. Vitamin E is the major chain-breaking antioxidant which protects membranes from lipid peroxidation (20). Vitamin E exists in biological membranes in a low molar ratio to unsaturated phospholipids, usually less than 0.1 nmol per mg of membrane protein, or, in other words, one molecule per 1000 to 2000 membrane phospholipid molecules, which are the main target of oxidation in membranes. Lipid peroxyl radicals can be generated in membranes at the rate of 1-5 nmol per mg of membrane protein per min., yet destructive oxidation of membrane lipids does not normally occur, nor is vitamin E rapidly depleted. Furthermore, deficiency states for vitamin E are remarkably difficult to induce in adult animals. These apparent paradoxes can be explained by "vitamin E recycling", in which the antioxidant ability of vitamin E is continuously restored by other antioxidants. Those antioxidants which recycle vitamin E are vitamin C, ubiquinols, and thiols (21, 22) .

Evidence for vitamin E recycling by DHLA has come from a number of studies. DHLA protects against microsomal lipid peroxidation, but only in the presence of vitamin E (23). DHLA prolonged the lag phase prior to the onset of low-level chemiluminescence, loss of vitamin E, and the rapid accumulation of thiobarbituric acid reactive substances in normal, but not in vitamin E-deficient, microsomes. α-Lipoic acid was not effective in either system. DHLA could have exerted its effect in this system by directly reducing tocopheroxyl radical, or by reducing other antioxidants (such as ascorbate) which then regenerated vitamin E. There may be a weak direct interaction between DHLA and the tocopheroxyl radical. The presence of DHLA reduces the tocopheroxyl radical ESR signal in liposomes exposed to UV light (our unpublished data). The use of UV light, which directly produces tocopheroxyl radicals, eliminate the possibility of iron chelation or other antioxidant effects of DHLA, and the use of liposomes eliminates the possibility of DHLA acting through other antioxidants. Therefore, in this system it seems likely that DHLA, which partitions mainly in the aqueous phase, directly reduces tocopheroxyl radicals at the membrane/water interface. However, the effect is weak, and the major recycling of vitamin E by DHLA in biological systems probably occurs through other antioxidants.

Bast and Haenen (24) have proposed that DHLA prevents lipid peroxidation by reducing glutathione disulfide, which in turn recycles vitamin E. This proposal is based on the observation that the combination of DHLA and glutathione disulfide, but not DHLA alone, prevented lipid peroxidation induced by Fe^{2+}/ascorbate (10). Alternatively, Kagan et al. (16) proposed that DHLA protects membranes against oxidation by recycling ascorbate, which in turn recycles vitamin E. Their ESR studies indicated DHLA-mediated reduction of ascorbyl radicals generated in the course of ascorbate oxidation by chromanoxyl (vitamin E or short-chain vitamin E homologues) radicals in DOPC liposomes and also showed that DHLA interacts with NADPH- or NADH-dependent electron transport chains to recycle vitamin E. Ascorbate-dependent recycling of vitamin E by DHLA has also been observed in human low density lipoproteins (25) and erythrocyte membranes (26). Additionally, there is evidence that α-lipoic acid administration *in vivo* can increase the level of ubiquinol in the face of oxidative stress (27) and ubiquinol is known to recycle vitamin E (28). Hence, current evidence points to the contention that DHLA can recycle vitamin E via glutathione, vitamin C, or ubiquinol, but the relative contributions of each avenue are not well-defined.

In fact, protection of other antioxidants by α–lipoic was suggested as early as 1959, by Rosenberg and Culik (29), who stated, with startling prescience, "*α–Lipoic acid, and even more so its dihydro derivative into which it is converted rapidly after entering cellular metabolism, might act as an antioxidant for ascorbic acid and tocopherols.*" In the studies of Rosenberg and Culik, α-lipoic acid was found to prevent symptoms of both vitamin E and vitamin C deficiency. Recently, we have found similar protective effects of α–lipoic acid administration in tocopherol-deficient hairless mice (30). Such results are consistent with a recycling of tocopherol and/or ascorbate by α–lipoic acid, but could also be explained by the ability of α-lipoic acid to spare ascorbate and vitamin E through its separate but overlapping radical-scavenging effects.

α–Lipoic acid also causes an increase in intracellular glutathione. Peinadao et al. (1989) showed that thiols appeared in the perfusate when α-lipoate was infused into the liver, but the exact nature of the thiols was not identified. Busse et al (31) added α–lipoic acid to murine neuroblastoma and melanoma cell lines, and observed a dose-dependent increase in GSH content of 30-70% compared to untreated controls. These investigators found similar elevations in lung, liver and kidney cells of mice injected (IP) daily with doses of 4, 8 or 16 mg/kg α–lipoic acid for 11 days. These results have been confirmed in human Jurkat cell lines, in which intracellular concentrations of GSH increase approximately 50% in 5 hours after addition of α–lipoic acid to the culture medium (32). Such elevations in GSH cannot be explained by reduction of GSSG, since GSSG is normally present at less than 10% the concentration of GSH (33). Following addition to the cell culture medium, α-lipoate is taken up by the cells and reduced to dihydrolipoate (34). Intracellular dihydrolipoate is rapidly released to the culture medium rich in cystine where this amino acid is reduced to the precursor amino acid for glutathione synthesis, cysteine. Extracellular reduction of cystine to cysteine and the rapid entry of the latter to the cell to stimulate glutathione synthesis may be one possible way of intracellular glutathione increase in response to α-lipoate treatment.

III. Effects against oxidatively induced injury

A. In vitro diabetic cataractogenesis

Rat lenses plus capsule were incubated in Modified Eagles Medium (MEM), MEM + 55.6 mM glucose, MEM + glucose + racemic LA (1mM), MEM + glucose + R-LA (1mM), or MEM + glucose + L-LA (1mM). Lactate dehydrogenase (LDH) leakage into the medium, a marker for decreased cell membrane integrity, was measured daily, as was lens opacity (Kilic et al., 1994). Lenses incubated in MEM alone showed no opacities and no LDH leakage; lenses incubated with high glucose started to develop opacities within 48 hours and LDH levels reached 5.49 U/ml/mg lens protein at that point. In contrast, lenses incubated with glucose + R-LA did not develop opacities or LDH leakage during the incubation period. Lenses incubated with glucose + S-LA developed opacities and LDH leakage at the same rate as those in high glucose alone, and racemic LA had little effect on either opacities or LDH leakage. In this system the protective effect was very clearly present only with the R-enantiomer for LA. S-lipoic acid was ineffective and the racemic mixture provided only slight protection.

B. Glutathione-deficient newborn rats

Administration of buthionine sulfoxide (BSO), a γ-glutamylcysteine synthetase inhibitor, to pre-weanling rats depletes glutathione and causes cataract formation. 100% of rats treated with BSO alone developed cataract, whereas only 40% of rats treated with 25 mg/kg racemic LA (injected IP) at the same time as BSO injection developed cataract; the R-enantiomer alone at this dose also partially prevented cataract, but not as effectively as the racemic mixture (Maitra et al. 1994). In the BSO-treated rats, levels of glutathione, ascorbate, and vitamin E dropped 56%, 55%, and 70%, respectively, but treatment with racemic LA in addition to BSO almost completely protected these antioxidants. In this system, in contrast to the in vitro cataractogenesis system described above, lipoic acid was administered intraperitoneally and had to travel to the lens, and cataract was induced through oxidative stress induced by glutathione deficiency, as opposed to induction by exposure to glucose. These and other factors may account for the difference in the effects of the enantiomers of LA in the two systems—in the in vivo system the racemic mixture was actually superior to R enantiomer alone.

C. Cerebral ischemia-reperfusion

In a cell culture model of hypoxia, cultured neurons (primary neuronal cultures derived form 7-day-old chick embryo telencephalon), are exposed to cyanide, followed by washout (35). In control hypoxic cultures, protein and ATP concentrations decreased due to neuronal death, but in dihydrolipoate-treated hypoxic cultures, protein concentration and ATP concentration were maintained at normal levels, indicating protection from hypoxic damage (35) . α−Lipoate was ineffective in this system if added one hour before hypoxia, but was protective if added 24 hours before; both the R- and the S- enantiomer were equally effective (36). These results indicate that α−lipoate is probably being reduced to dihydrolipoate in these cells, and that pathways of reduction which are specific for both the R- and the S-enantiomers operate in neuronal cells. In our studies we have found that the S-enantiomer is the preferred substrate of glutathione reductase, present in the cytosol, while the R-enantiomer (the natural form present in the E2 enzyme of α-keto acid dehydrogenase complexes) is preferred by dihydrolipoamide dehydrogenase (the E3 enzyme), in mitochondria (36). Thus, it seems likely that both mitochondrial and cytosolic pathways are responsible for reduction of α–lipoate in neuronal cells to its more potent antioxidant form, dihydrolipoate.

Neuroprotective effects have also been found in animal models. Prehn et al., in a study with middle cerebral artery occlusion in mice, demonstrated that in this model of focal ischemia, treatment with dihydrolipoate, but not α-lipoate, reduced the size of the infarct (35, 37) . Backhaus et al., using the same model, also found a protective effect of dihydrolipoate but not α–lipoate (38). In an investigation of the mouse model of focal cerebral ischemia, Wolz and Krieglstein found that α–lipoate administered subcutaneously was protective, but α–lipoate administered either intraperitoneally or intracisternally was not (39). The authors speculate that, at least in the case of intraperitoneal administration, α–lipoate was metabolized in the liver so that brain levels remained low and no protective effect was seen. The R- and S- enantiomers of α–lipoate were also tested in this study; both were found to be effective if administered 2 hours before induction of ischemia, but the S- enantiomer was more effective than the R-enantiomer if administered only one hour before ischemia. These results confirm studies with cells in tissue culture such as human lymphocytes and keratinocytes indicating that both glutathione reductase and dihydrolipoamide dehydrogenase pathways of α–lipoate reduction operate in vivo, although it is possible that reduction is occurring at some other location than brain.

Cao and Phillis also observed a protective effect of α–lipoate against ischemia-reperfusion injury in the Mongolian gerbil model, which offers the advantage of more complete ischemia (40). Gerbils treated with α–lipoate for 7 days prior to ischemia-reperfusion exhibited less change in locomotor activity and less damage to the CA1 hypocampal pyramidal cell layer, compared to saline-treated controls.

The work of Panigrahi et al. (41)illustrates the necessity of the choice of an appropriate antioxidant for treatment of brain disorders. In rats subjected to reperfusion following ischemia induced by bilateral carotid artery occlusion, pretreatment with GSH isopropyl ester had no protective effect, consistent with the studies of Anderson et al.(42), in which GSH levels in brain were not increased by administration of glutathione esters. In contrast, pretreatment with α–lipoate (25 mg/kg given intravenously immediately before ischemia) reduced mortality from 78% to 26% in the 24 hours following reperfusion. This represents the most remarkable effect of a therapeutic substance in protection against cerebral ischemia-reperfusion so far known. Lipoate pretreatment also almost completely abolished ischemia/reperfusion-induced losses of glutathione in the cortex, striatum, and hippocampus. α–Lipoate pretreatment dramatically decreased lipid peroxidation in these brain areas also; animals pretreated with α–lipoate experienced a 60% average rise in thiobarbituric acid reactive substances (TBARS) while untreated animals experienced a 225% average rise in TBARS. Hence in this study α–lipoate treatment was shown to be effective in preventing antioxidant loss, attenuating oxidative damage, and dramatically increasing survival after cerebral ischemia-reperfusion.

IV. Thiol redox regulation

A. Introduction

Thiols influence the intracellular and extracellular redox balance. Low molecular weight thiols interact with protein targets affecting enzyme activity, transcription, and gene expression. Glutathione, the most abundant low molecular weight thiol in cells, acts as a metabolic antioxidant and redox regulator in a variety of ways. It is a substrate for glutathione-dependent enzymes, such as transhydrogenases and GSH peroxidases and interacts with other antioxidants, regenerating dehydroascorbate to ascorbic acid. However, GSH does not reduce thioredoxin, a key thiol regulator of cell function, whereas dihydrolipoate readily reduces thioredoxin.

In response to stresses such as injury and infection, organisms must rapidly marshal a host of responses. Various genes must be activated, and a central, coordinating factor in this activation of rapid response gene is nuclear factor κB (NF-κB). In addition, NF-κB plays a crucial role in modulating gene expression during growth and development. In pathological conditions such as viral infection, NF-κB could be used by the invading pathogen to assist in viral gene expression. Hence, the modulation of NF-κB has far-reaching significance for a variety of pathological conditions in which inflammation, growth, or viral activation occur, such as atherosclerosis and HIV infection.

NF-κB is actually a family of Rel proteins, transcription factors that share common characteristics. They consist of hetero- or homodimers in association with an inhibitory protein, IκB. Phosphorylation and dissociation of IκB from Rel protein dimers, followed by its degradation allows the dimeric DNA-binding NF-κB to enter the nucleus and regulate gene expression **(Fig. 1)** (43, 44).

A host of genes have been shown to be modulated by NF-κB. These include genes for cytokines and growth factors, immunoreceptors, adhesion molecules, acute phase proteins, transcription factors and regulators, NO-synthase, and viral genes (44).

Role of redox balance in modulating NF-κB activation

There are several lines of evidence that reactive oxygen species (ROS) may be the final common signal for a number of stimuli that activate NF-κB(43) (Brown et al. 1993, Sen and Packer 1996).

First, many of the stimuli which activate NF-κB are also known to cause intracellular increases in ROS. These include TNF-α, IL-1, phorbol myristate acetate (PMA), lipopolysaccharide (LPS), UV light, and gamma irradiation (45-48).

Second, administration of hydrogen peroxide can directly stimulate NF-κB activation (45, 49). However, several other ROS (hydroxyl radical, singlet oxygen etc.) have not been effective (44)

Third, a diverse array of antioxidants can block NF-κB activation. These include N-acetyl-L-cysteine (50), dithiocarbamates (46, 50), vitamin E derivatives (51, 52), catechol derivatives (53), anethole dithiolthione (54) and α–lipoate (55-59). Conversely, in certain instances agents which deplete cellular thiol antioxidants have been observed to induce NF-κB activation (60)

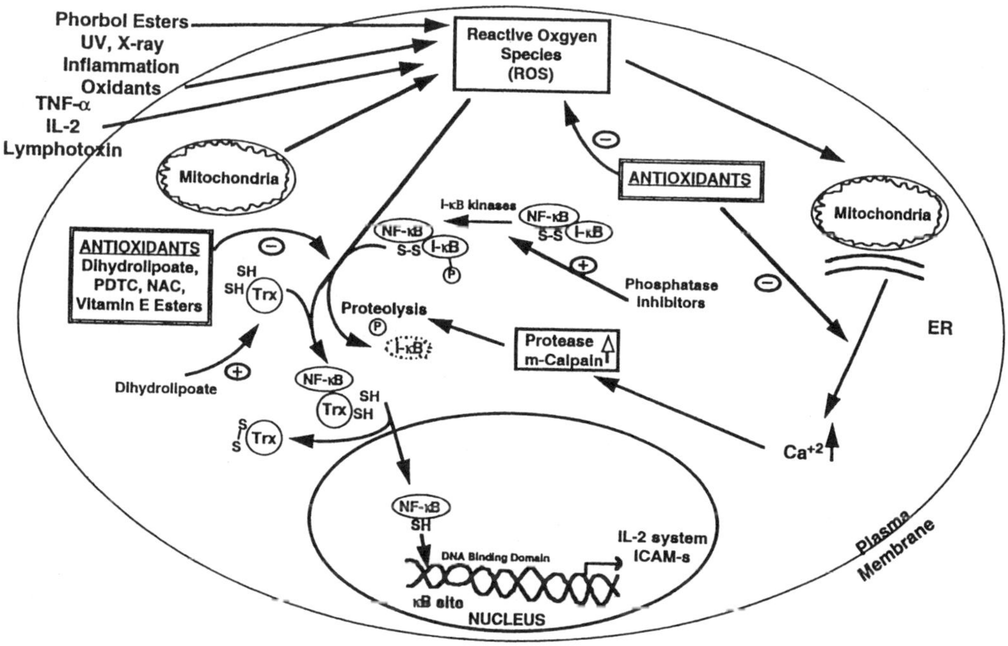

Fig. 1 Oxidant induced regulation of signal transduction

Effects of lipoate and dihydrolipoate on NF-κB

A number of researchers have reported antioxidants, including ebselen, 2-mercaptoethanol, glutathione, PDTC, diethyldithiocarbamate, disulfiram, orthophenanthroline, anethole dithiolthione, deferroxamine, NDGA, vitamin C, L-cysteine, and BHA to inhibit NF-κB activation (45, 54). We have shown that vitamin E derivatives are potent suppressors of NF-κB activation (51, 52). Incubation of human Jurkat T cells with vitamin E acetate or α–tocopheryl succinate exhibited a concentration dependent inhibition of NF-κB activation. α–Tocopherol or succinate at these concentrations had no apparent side effects. Also 2,2,5,7,8-pentamethyl-6-hydroxychromanol (PMC) was extremely effective, causing complete inhibition of NF-κB activation at 10 μM. Therefore, it was of interest to us to examine the effects of α–lipoate on NF-κB activation.

Suzuki et al. 1992 observed that incubation of Jurkat T cells with mM concentration of lipoate prior to the stimulation with TNF-α or phorbol 12-myristate 13-acetate inhibited NF-κB activation in a dose-dependent manner, with 4 mM required for complete inhibition. α–Lipoate at these concentrations did not affect either cell viability or oct-1 DNA binding activity. Direct supplementation of DHLA in the incubation medium also inhibited the NF-κB activation induced by TNF-α (57). Although the potency of DHLA on NF-κB inhibition was comparable to that of α–lipoic acid, millimolar concentrations of DHLA affected cell viability. Both R- and S-enantiomers of lipoate are effective in inhibiting NF-κB activation simulated by TNF-α treatment in Jurkat T cells after 2 hrs of preincubation (57). DNA binding of NF-κB, which occurs after the activation and involves phosphorylation, dissociation and degradation of the IκB subunit. DHLA enhanced DNA binding of NF-κB, while α–lipoate inhibited (58). Hence, there is an apparent paradox in the redox regulation of NF-κB. The observations from this study suggest that the α–lipoate action preventing the NF-κB DNA binding activity is, at least in part, the underlying mechanism for the observed inhibitory action against NF-κB activity. A dual action of dihydrolipoate, by inhibiting the TNFα–induced NF-κB activation and potentiating the DNA binding of activated NF-κB, suggests two modes of redox regulation indeed exist in cell signaling for NF-κB: first, a requirement of oxidative processes in activating NF-κB; and second, a requirement of reductive processes in binding to DNA.

Because α-lipoate can increase intracellular glutathione levels (31, 32), it is possible that the α-lipoate induced inhibition of NF-κB was actually mediated by glutathione. To test this, Wurzburg T-cells (a Jurkat T-cell sub-clone) were pre-incubated with BSO (0.2 mM) either in the presence or absence of 0.5 mM α-lipoate for 18 h. Such BSO treatment decreased intracellular GSH levels by 80-90%. Although α-lipoate treatment can markedly increase intracellular glutathione levels, this effect is arrested in cells cotreated with BSO and α-lipoate. Even in glutathione synthesis-arrested cells, α-lipoate inhibited NF-κB activation induced by H_2O_2, TNFα or PMA, as assessed by EMSA (61). This suggests that the transcriptional regulation by lipoate is independent of its effect on cell glutathione, perhaps because the reduced form (DHLA) acts as a dithiol reductase.

Studies on the in vitro DNA binding assay have shown that activated nuclear extracts of NF-κB are prevented from binding to the NF-κB consensus site by lipoate. Also, diamide inhibits the DNA binding step; this can be overcome by adding dihydrolipoate. This effect may be mediated directly by dihydrolipoate reducing critical thiols on the activated transcription factor, or indirectly by affecting thioredoxin, which has been shown to be necessary in this step.

Expression of down-stream products of NF-κB activation have also been investigated, and PMA-induced IL-2 receptor expression and H_2O_2-induced ICAM 1 expression in cultured Jurkat T-cells are inhibited by lipoate pretreatment for 18 h (S. Roy, C.K. Sen, unpublished observations).

B. Effects on intracellular calcium.

Intracellular calcium storage organelles sequester calcium and release it via receptor operated calcium channels (62). Proteins containing PEST regions are susceptible to degradation by calcium-dependent intracellular proteases, and IkB is known to be one such protein. Hence, we tested the hypothesis that oxidant induced increases in $[Ca^{2+}]_i$ is implicated in NF-κB activation. We used human Jurkat T-cells, which are not highly sensitive to oxidant (H_2O_2) induced NF-κB activation (63) as well as the Wurzburg T-cells, which are highly sensitive to such activation. We sought to show if an oxidant induced sustained increase in $[Ca^{2+}]_i$ is a crucial factor in the NF-κB activation process, and to determine whether α-lipoate exerts its NF-κB regulatory effect, at least in part, by stabilizing the elevation of $[Ca^{2+}]_i$ following oxidant challenge (64).

Real time flow-cytometric monitoring of oxidant induced changes in $[Ca^{2+}]_i$ in a gated morphometrically homogeneous Jurkat T-cell population showed that within 60-90s of 0.25 mM H_2O_2 challenge $[Ca^{2+}]_i$ was remarkably elevated, but in less than 15-20 min after oxidant challenge, $[Ca^{2+}]_i$ returned to baseline levels. Pretreatment of the cells with α-lipoate partially stabilized oxidant induced perturbation of intracellular calcium homeostasis, and with higher concentrations (2 mM) of lipoate pretreatment, oxidant induced elevation of $[Ca^{2+}]_i$ was suppressed to the point that oxidant treatment had only a minimal effect on $[Ca^{2+}]_i$. The kinetics of oxidant dependent changes in $[Ca^{2+}]_i$ in Wurzburg cells was quite different compared to the parental Jurkat cell line: in Wurzburg T-cells, the 0.25 mM H_2O_2 induced peak $[Ca^{2+}]_i$ was observed after about 10 min of the challenge, and even after 60 min of oxidant treatment, $[Ca^{2+}]_i$ was higher than baseline. Pretreatment of Wurzburg cells with α-lipoate resulted in lower post-oxidant $[Ca^{2+}]_i$ mobilization.

When nuclear protein extracts of H_2O_2 stimulated Wurzburg T-cells were examined by EMSA using the labeled NF-κB binding oligonucleotide, a distinct NF-κB retardation band was observed. Pretreatment of the Wurzburg cells with α-lipoate resulted in a marked inhibition of H_2O_2 induced NF-κB activation in these cells. In addition, H_2O_2 induced activation of NF-κB was completely inhibited by loading the cells with EGTA-AM, a calcium chelator which binds intracellular calcium. In contrast, Jurkat T-cells were barely responsive to H_2O_2 with respect to NF-κB activation. In Jurkat T-cells treated with thapsigargin, the sarco-endoplasmic reticulum membrane calcium pump inhibitor, the $[Ca^{2+}]_i$ was remarkably elevated but NF-κB activation was only weakly induced. However, H_2O_2 potentiated the thapsigargin induced NF-κB activation in Wurzburg cells. The inhibition by EGTA-AM treatment indicates that the presence of intracellular Ca^{2+} is a necessary condition for NF-κB activation. In this context, the inhibitory effect of α-lipoate on intracellular calcium levels is probably a component of its overall effect on reactive oxygen species, redox effects on cell thiols and early steps in NF-κB activation.

V. Conclusion

α-Lipoic acid affects intracellular redox status by direct radical scavenging and by maintaining reduced forms of antioxidants in the antioxidant network and by affecting the redox status of protein thiols. The major human T cell regulatory effects of α–lipoic acid are: i) inhibition of NF-κB activation in response to a wide variety of oxidant-induced stimuli, ii) decreased oxidant induced perturbation of intracellular Ca^{2+}, iii) increases in cellular glutathione levels, and iv) decreases in oxidant-induced T-cell gene expression such as adhesion molecules. These properties may have clinical significance for reducing oxidative-stress dependent diseases (atherosclerosis, diabetes, diabetic polyneuropathy, neurodegeneration, viral activation, etc.).

References

1. Suzuki, YJ, Tsuchiya, M, and Packer, L.. *Free Rad. Res. Comms.* (1991) **15**: 255-263.
2. Scott, BC, Aruoma, OI, Evans, PJ, O'Neill, C, van der Vliet, A, Cross, CE, Tritschler, H, and Halliwell, B. *Free Rad. Res.* (1994) **20**: 119-133.
3. Matsugo, S, Yan, L-J, Han, D, Tritschler, HJ, and Packer, L. *Biochem. Biophys. Res. Commun.* (1995) **208**: 161-167.
4. Haenen, GR and Bast, A. *Biochem. Pharmacol.* (1991) **42**: 2244-2246.
5. Stevens, B, Perez, SR, and Small, RD. *Photochem. Photobiol.* (1974) **19**: 315-316.
6. Stary, FE, Jindal, SJ, and Murray, RW. *J. Org. Chem.* (1975) **40**: 58-62.
7. Kaiser, S, Di Mascio, P, and Sies, H, *Lipoat und Singulettsauerstoff*, in *Thioctsäure*, Borbe, HO and Ulrich, H, Editors. 1989, pmi Verlag GmbH: Frankfurt. p. 69-76.
8. Devasagayam, TPA, Subramanian, M, Pradhan, DS, and Sies, H. *Chem.-Biol. Interactions* (1993) **86**: 79-92.
9. Müller, L and Menzel, H. *Biochim. Biophys. Acta* (1990) **1052**: 386-391.
10. Bast, A and Haenen, GRMM. *Biochim. Biophys. Acta.* (1988)**963**:558-561.
11. Sigel, H, Prijs, B, McCormick, DB, and Shih, JCH. *Arch. Biochem. Biophys.* (1978) **187**: 208-214.
12. Ou, P, Tritschler, H, and Wolff, SP. *Biochem.Pharmacol.* (1995) **50**: 123-126.
13. Searls, RL and Sanadi, DR. *J. Biol. Chem.* (1960) **235**: 2485-2491.
14. Scott, I, Duncan, W, and Ekstrand, V. *J. Biol. Chem.* (1963) **238**: 3928-3939.
15. Jocelyn, PC. *Eur. J. Biochem.* (1967) **2**: 327-331.
16. Kagan, VE, Shvedova, A, Serbinova, E, Khan, S, Swanson, C, Powell, R, and Packer, L. *Biochem. Pharmacol.* (1992) **44**: 1637-1649.
17. Matsugo, S, Yan, L-J, Han, D, Tritschler, HJ, and Packer, L. *Biochem. Mol. Biol. Int.* (1995) **37**: 375-383.
18. Suzuki, YJ, Tsuchiya, M, and Packer, L. *Free Rad. Res. Comms.* (1993) **18**: 115-122.
19. Matsugo, S, Konishi, T, Matsuo, D, Tritschler, HJ, and Packer, L. *Biochem Biophys Res Commun.* (1996) *in press*:
20. Burton, GW and Ingold, KU. *J. Am. Chem. Soc.* (1981) **103**: 6472-6477.
21. Sies, H. *Eur. J. Biochem.* (1993) **215**: 213-219.
22. Packer, L. *Lipid-Soluble Antioxidants: Biochemistry and Clinical Applications*, Ong, ASH and Packer, L, Editors. 1992, Birkhauser Verlag: Boston. p. 1-16.
23. Scholich, H, Murphy, ME, and Sies, H. *Biochim. Biophys. Acta.* (1989) **1001**: 256-261.
24. Bast, A and Haenen, GRMM. *Therapy and Preventive Medicine*, Emerit, I, Packer, L, and Auclair, C, Editors. 1990, Plenum Press: New York. p. 111-116.
25. Kagan, VE, Serbinova, EA, Forte, T, Scita, G, and Packer, L. *J. Lipid Res.* (1992) **33**: 385-397.
26. Constantinescu, A, Han, D, and Packer, L. *J Biol Chem* (1993) **268**: 10906-10913.

27. Gotz, ME, Dirr, A, Burger, R, Wanetzky, B, Weinmuller, M, Chan, WW, Chen, SC, Reichmann, H, Rausch, WD, and Riederer, P. *Eur. J. Pharmacol.* (1994) **266**: 291-300.
28. Kagan, V, Serbinova, E, and Packer, L. *Bioch. Biophys. Res. Commun.* (1990) **169**: 851-857.
29. Rosenberg, HR and Culik, R. *Arch. Biochem. Biophys.* (1959) **80**: 86-93.
30. Podda, M, Tritschler, HJ, Ulrich, H, and Packer, L. *Biochem. Biophys. Res. Commun.* (1994) **204**: 98-104.
31. Busse, E, Zimmer, G, Schopohl, B, and Kornhuber, B. *Arzneimittel-Forschung* (1992) **42**:829-831.
32. Han, D, Tritschler, HJ, and Packer, L. *Bioch. Biophys. Res.* (1995) **207**: 258-264.
33. Halliwell, B and Gutteridge, JMC, *Free radicals in biology and medicine, second edition*. 1989, Oxford, England: Clarendon Press.
34. Handelman, GJ, Han, D, Tritschler, H, and Packer, L. *Biochem. Pharmacol.* (1994) **47**: 1725-1730.
35. Prehn, JHM, Karkoutly, C, Nuglisch, J, Peruche, B, and Krieglstein, J. *J. Cereb. Blood Flow Metab.* (1992) **12**: 78-87.
36. Muller, U and Krieglstein, J. *J. Cereb. Blood Flow Metab.* (1995) **15**: 624-630.
37. Prehn, JH, Peruche, B, Karkoutly, C, Roßberg, C, Mennel, HD, and Krieglstein, J. *Pharmacology of Cerebral Ischemia.*, Krieglstein, J and Oberpichler, H, Editors. 1990, Wissenschaftliche Verlagsgesellscheaft mbH: Stuttgart. p. 357-362.
38. Backhaus, C, Karkoutly, C, Welsch, M, and Krieglstein, J. *J. Pharmacol. Meth.* (1992) **27**: 27-32.
39. Wolz, P and Krieglstein, J. *J. Cereb. Blood Flow Metab.* (1995) **Submitted**:
40. Cao, X and Phillis, JW. *Free Rad. Res.* (1995) **23**: 365-370.
41. Panigrahi, M, Sadguna, Y, Shivakumar, BR, Kolluri, VR, Roy, S, Paker, L, and Ravindranath, V. (*in press*)
42. Anderson, A, Powrie, F, Puri, RN, and Meister, A. *Arch. Bioch. Biophys.* (1985) **239**: 538-548.
43. Brown, K, Park, S, Kanno, T, Franzoso, G, and Siebenlist, U. *Proc. Natl. Acad. Sci. USA* (1993) **90**: 2532-2536.
44. Siebenlist, U, Franzoso, G, and Brown, K. *Ann. Rev. Cell Biol.* (1994) **10**: 405-455.
45. Schreck, R, Albermann, K, and Baeuerle, PA. *Free Radical Res. Commun.* (1992) **17**: 221-237.
46. Schreck, R, Meier, B, Maennel, DN, Droge, W, and Baeuerle, A. *J. Exp. Med.* (1992) **175**: 1181-1194.
47. Geng, Y, Zhang, B, and Lotz, M. *J. Immunol.* (1993) **151**: 6692-6700.
48. Schieven, GL, Kirihara, JM, Myers, DE, Ledbetter, JA, and Uckun, FM. *Blood* (1993) **82**: 1212-1220.
49. Schreck, R, Rieber, P, and Baeuerle, P. *EMBO J* (1991) **10**. 2247-2258.

50. Meyer, R, Caselmann, WH, Schluter, V, Schreck, R, P.H., H, and P.A., B. *EMBO J.* (1992) **11**: 2992-3001.

51. Suzuki, YJ and Packer, L *Biochem. Biophys. Res. Commun.* (1993) **193**: 277-283.

52. Suzuki, YJ and Packer, L *Biochem. Mol. Biol. Intl.* (1994) **31**: 693-700.

53. Suzuki, YJ and Packer, L *Biochem. Mol. Bio. Intl.* (1994) **32**: 299-305.

54. Sen, CK, Traber, K, and Packer, L *Biochem. Biophys. Res. Commun.* (1996) **218**: 148-153.

55. Suzuki, YJ, Aggarwal, BB, and Packer, L. *Biochem. Biophys. Res. Commun.* (1992)**189**:1709-1715.

56. Packer, L and Suzuki, YJ. *Molec. Aspects Med.* (1993) **14**: 229-239.

57. Suzuki, YJ and Packer, L. *Biological oxidants and antioxidants*, Packer, L and Cadenas, E, Editors. 1994, Hippokrates Verlag: Stuttgart.

58. Suzuki, YJ, Mizuno, M, Tritschler, HJ, and Packer, L *Bioch. Mol. Biol. Int.* (1995) **36**: 241-246.

59. Bessho, R, Matsubara, K, Kubota, M, Kuwakado, K, Hirota, H, Wakazono, Y, Lin, YW, Okuda, A, Kawai, M, and Nishikomori, R. *Biochem. Pharmacol.* (1994) **48**: 1883-1889.

60. Staal, FJT, Roederer, M, Herzenberg, LA, and Herzenberg, LA *Proc. Natl. Acad. Sci. USA* (1990) **87**: 9943-9947.

61. Sen, CK and Packer, L. *Faseb J* (1996) **10**: 709-720.

62. Bootman, MD and Berridge, MJ. *Cell* (1995) **83**: 675-678.

63. Anderson, MT, Staal, FJT, Gitler, C, Herzenberg.L.A., and L.A., H. *Proc. Natl. Acad. Sci. USA* (1994) **91**: 11527-11531.

64. Sen, CK, Roy, S, and Packer, L *FEBS Letts* (1996) **385**: 58-62.

II

Stress-induced Signal Transduction Systems for Biodefence

5. Regulatory Mechanisms of Gene Activation Induced by DNA Damage

Mitsuru NENOI,[1)] Kazuei MITA[2)] and Sachiko ICHIMURA[1)]

[1)]Division of Biology and Oncology, [2)]Genome Research Group, NIRS, Chiba 263, Japan

Gene induction by DNA-damaging agents

Various kinds of environmental stress cause DNA damage. These include carcinogenic and anti-neoplastic chemical agents, ultraviolet lights in sunshine and ionizing radiations from natural as well as industrial sources. Damage of DNA is classified into several groups, including pyrimidine dimers, DNA strand scission, base alkylations and crosslinkages with either DNA molecules themselves or with other intracellular macromolecules. Responding to DNA damage, cells are known to induce numerous genes specifically or nonspecifically to the type of DNA damages. Induction of these genes would be an important process in biodefensive response at the cellular and individual levels.

Friedberg *et al.*[1)] have summarized the mammalian genes, which have been reported to be inducible by DNA-damaging agents at the mRNA level or the protein level. They include several genes known to play a defensive role, such as metallothionein, ornithine decarboxylase, O^6-methylguanine-DNA methyltransferase, N^3-methyladenine-DNA glycosylase and WAF1. However it is notable that they also include many genes encoding transcription factors, which regulate many DNA damage-inducible genes directly. For example, c-*fos*, c-*jun*, *jun*B are members of AP-1 family, and NFκB, EGR-1, c-*myc* and p53 are also transcription factors. Recent studies have revealed many aspects of regulation mechanisms and functions of these transcription factors in the context of transcriptional induction responding to DNA-damaging agents, especially for AP-1, NFκB and p53.

Responding to DNA-damaging agents, the earliest detectable alteration of the cells is hyperphosphorylation of the EGF receptor[2)] and activation of the Src tyrosine kinases at the plasma membrane.[3)] These responses take place within 5 minutes after treatment of cells with DNA-damaging agents. Activation of Src tyrosine kinases is followed by the activation of Ha-*ras* and Raf-1.[3)] Then the Raf-1 activates the MAP kinase[3)] and the JNKs[4)] to induce activation of AP-1 on one hand, and promote degradation of the inhibitory factor IκBα leading to translocation of the NFκB into the cell nucleus[5)] on the other. Both AP-1 and NFκB are also under the redox regulation mediated by the factor Ref-1.[6)] p53 is induced by the DNA-damaging agents probably mediated by the ATM, which is the product of the gene responsible for ataxia telengiectasia.[7)]

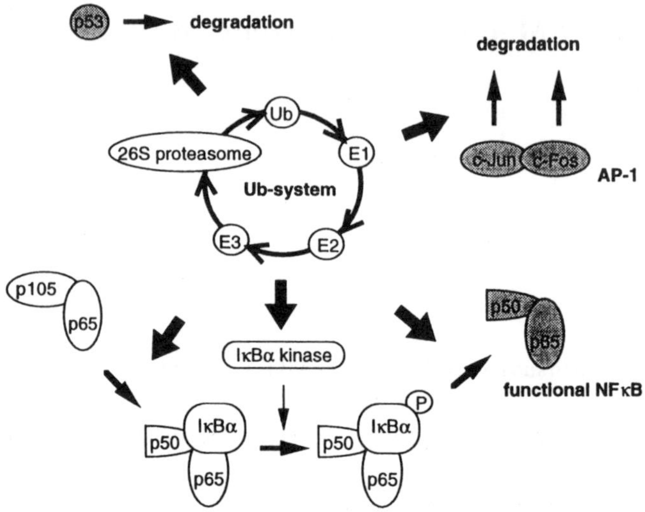

Fig. 1 The roles of the ubiquitin system in metabolizing the transcription factors AP-1, NFκB and p53. See text.

Ubiquitin system

There is accumulating evidence indicating that a highly conserved protein, ubiquitin, plays an essential role in metabolizing AP-1, NFκB and p53 (Figure 1). Ubiquitin is the key component of what is called the ubiquitin system. The ubiquitin system functions in degradation and modification of the numerous regulatory proteins.[8] The substrates of the ubiquitin system are specifically ubiquitinylated by the cooperation of E1, E2 and E3 enzymes, and are degraded or processed in the 26S proteasome. A significant role for ubiquitin in the regulation of the transcriptional induction by DNA-damaging agents is expected primarily because the transcription factors p53 and AP-1 are degraded by the ubiquitin system.[9,10] Moreover, it has been shown that the activation of NFκB requires the ubiquitin system at multiple steps, including the processing of the p105 precursor to produce the mature p50 subunit,[5] the activation of the IκBα kinase,[11] and degradation of the phosphorylated IκBα, resulting in production of the functional NFκB.[5]

We previously observed that the ubiquitin gene of HeLa cells is induced by ultraviolet light and a phorbol ester, TPA at the mRNA level.[12] In humans, ubiquitin is encoded by the multiple ubiquitin gene family composed of UbA, UbB and UbC. Among these genes, the UbC gene is induced by UV light dose-dependently and transiently with the maximum induction at about 3 hours after irradiation. The UbC gene is also induced by TPA with very similar kinetics. The induction of the UbC gene was also observed after irradiation with a high LET radiation of the carbon beam.

Based on these observations, we hypothesized that ubiquitin and transcription factors mutually regulate each other, and control the induction level of subsequent defensive gene products. In order to prove this hypothesis, it is necessary to examine which signals the UbC gene receives from the signaling cascade triggered by DNA-damaging agents. More directly, the question is "Which transcription factor controls the UbC gene?" Then we analyzed the upstream control region of the UbC gene.

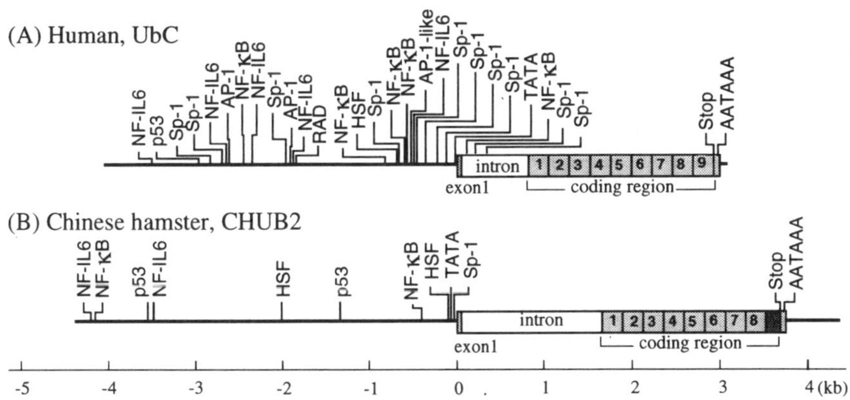

Fig.2 Schematic representation of the overall structure of the UbC and CHUB2 genes. Sites of a perfect match to the consensus sequences for transcription factors AP-1, NFκB, p53, SP-1, NF-IL6, HSF and RAD sequence in the upstream region and around the transcription initiation site of each gene are indicated, together with the TATA box. The numbers given in the hatched coding region boxes indicate the individually encoded ubiquitins. The darkly shaded box in the 3' coding region of the CHUB2 gene represents the non-ubiquitin polypeptide of 50 a.a.

Structure of the upstream region of the ubiquitin gene

Figure 2(A) shows the whole structure of the UbC gene encoding the tandemly repeated 9 ubiquitins.[13] Based on the data base of the recognition sequence for transcription factors, the site of AP-1, NFκB, p53 as well as NF-IL6, SP-1, RAD factor and the HSF were plotted. Interestingly, recognition sequences for these factors were distributed in two clusters. In addition, it is important to note that there are many SP-1 sites in the upstream region of the UbC gene because the DNA-bound SP-1 is known to be able to self-associate, bringing together distant DNA segments.[14] Figure 2(B) shows the whole structure of the Chinese hamster ubiquitin gene, CHUB2. This gene is considered to be the Chinese hamster homologue of the human UbC gene because of the high homology of the nucleotide sequence in the 5' and the 3' untranslated regions.[15] However, we could not observe the induction of the CHUB2 gene by UV light.[13] Obviously the upstream region of the CHUB2 gene is very distinct from that of the UbC gene. Thus it is not characterized by clustered recognition sequences for transcription factors as seen in the UbC gene. There are no AP-1 sites, very few NFκB sites and even fewer SP-1 sites, reflecting the low GC content of this region compared with the corresponding region of the human UbC gene. From this result, It is considered likely that the clustered distribution of transcription factor binding sequences in the control region and the presence of a lot of SP-1 sites in this region may be necessary, at least in part, for the UbC gene induction by UV light.

In order to determine the factors and the sequence elements actually regulating the induction of the UbC gene, we carried out a genomic footprinting experiment using the DNase I as the DNA-attacking agent. Figure 3 shows the designing of the primers for the genomic footprinting experiment. Four independent sets of primers were designed to investigate the footprinting pattern on both strands. The regions examined here contain only the AP-1 sites, and the regions containing the NFκB and p53 sites have not yet been examined.

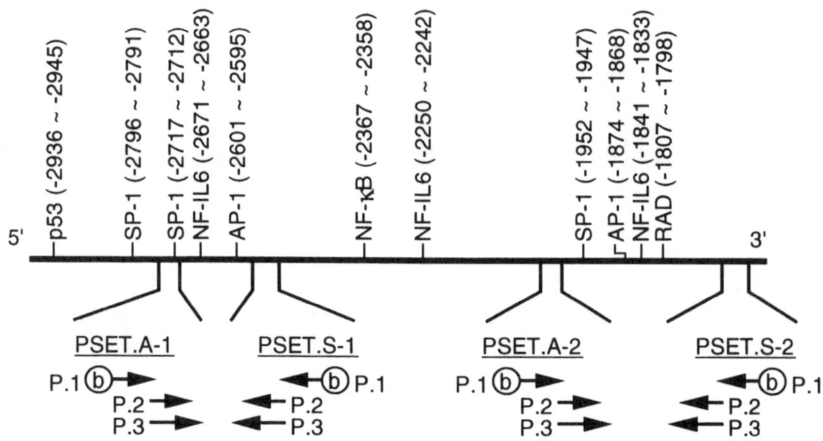

Fig.3 The design of primers for the genomic footprinting experiments. The annealing sites and orientations of the primers are illustrated on the map of recognition sites for transcription factors. The circled b represents biotin attached to the 5' end of the primer 1 (P.1).

Figure 4 shows the footprinting pattern around the AP-1 site at 2.6 kb upstream from the transcription initiation site. At the AP-1 site of -2601 bp to -2595 bp, a clear footprint was observed on the anti-sense strand even before transcriptional enhancement by UV light. In addition, at the NF-IL6 site of -2671 bp to -2663 bp on both strands, a footprint was also observed, as well as one at the SP-1 site of -2717 bp to -2712 bp on the sense-strand. These observations imply that the enhanced expression of the UbC gene after UV irradiation is not primarily brought about by a *de novo* binding of either of these factors to the unoccupied cognate sequences of DNA. However, on the sense-strand, an enhancement of protection around the NF-IL6 site and an appearance of protection just downstream of the SP-1 site was observed after UV irradiation. These observations indicate that minor alterations of the chromatin structure were generated by UV irradiation around the recognition sites of these factors. Furthermore, a slight and uniform sensitization to DNase I attack over the region between the AP-1 site through the NF-IL6 site on the anti-sense strand was observed after UV irradiation, which, it might be speculated, may result from an extensive alteration of the chromatin structure caused by some interaction between AP-1 and NF-IL6 bound to their cognate sequences of DNA.

Figure 5 shows the footprinting pattern around the AP-1 site at about 1.9 kb upstream from the transcription initiation site. At the AP-1 site of -1874 bp to -1868 bp, a footprint was again clearly observed on both strands even before UV irradiation, while the footprint on the anti-sense strand was relatively enhanced after UV irradiation. At the region of -1841 bp to -1833 bp, which is between the AP-1 site and the NF-IL6 site, protection was observed on both strands even before UV irradiation, and this protection was also relatively enhanced on the anti-sense strand after UV irradiation. At the RAD sequence of -1807 bp to -1798 bp, a clear footprint was observed on both strands before UV irradiation. The downstream neighboring region of the RAD sequence (-1793 bp to -1800 bp) on the sense strand became somewhat sensitized after irradiation.

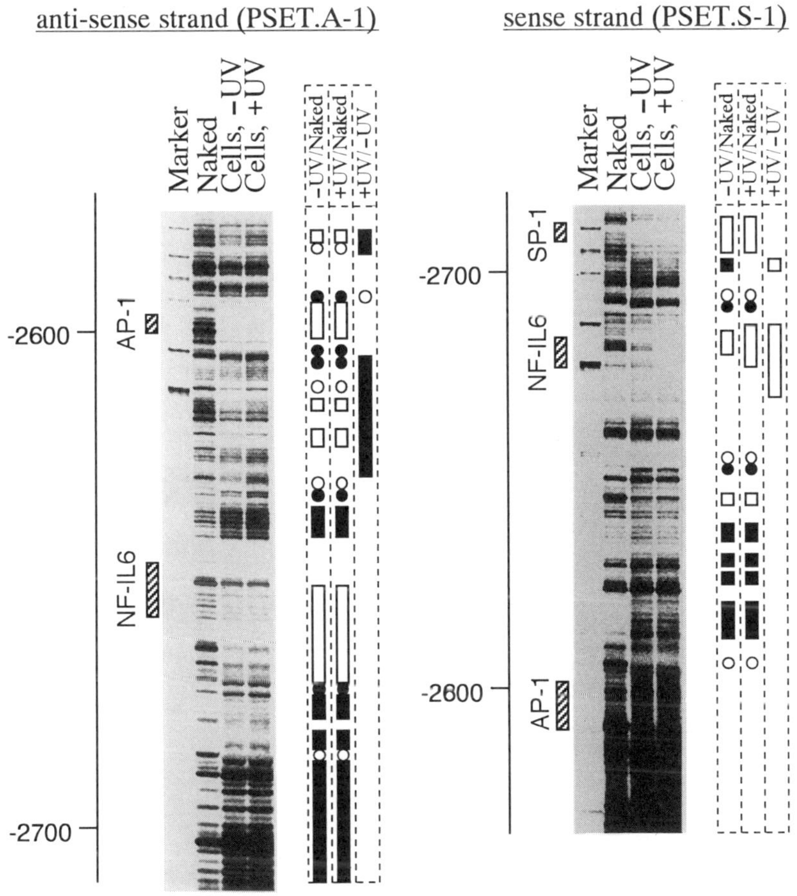

Fig.4 Genomic footprinting analysis in the region from -2750 bp to -2550 bp. The left panel, indicated as PSET.A-1, is the autoradiogram showing the footprinting pattern on the anti-sense strand, and the right panel, indicated as PSET.S-1, is on the sense strand. The scale in bp represents the distance from the transcription initiation site. The locations of the recognition sequences for transcription factors are indicated by hatched boxes immediately to the right of the scale. The lane designated [Marker] shows the plasmid pBR322 digested with *Msp*I. The lanes designated [Naked], [Cells, -UV] and [Cells, +UV] show the products of the LMPCR performed with templates of naked DNA digested with DNase I *in vitro*, nuclear DNA digested with DNase I in unirradiated cells *in vivo*, and nuclear DNA digested with DNase I in cells irradiated with 20J/m^2 of UV light *in vivo*, respectively. Comparisons of the chromatin sensitivity to DNase I attack are shown on the right hand side of the autoradiograms. [-UV/Naked] and [+UV/Naked] represent the sensitization or protection of chromatin in unirradiated and UV-irradiated cells, respectively, compared with naked DNA. [+UV/-UV] represents the sensitization or protection of the chromatin in UV-irradiated cells compared with that in unirradiated cells, and therefore reflects the alteration of the chromatin structure resulting from UV irradiation. Sensitized regions and sites are indicated by closed bars and closed circles, respectively, and protected regions and sites are indicated by open bars and open circles, respectively.

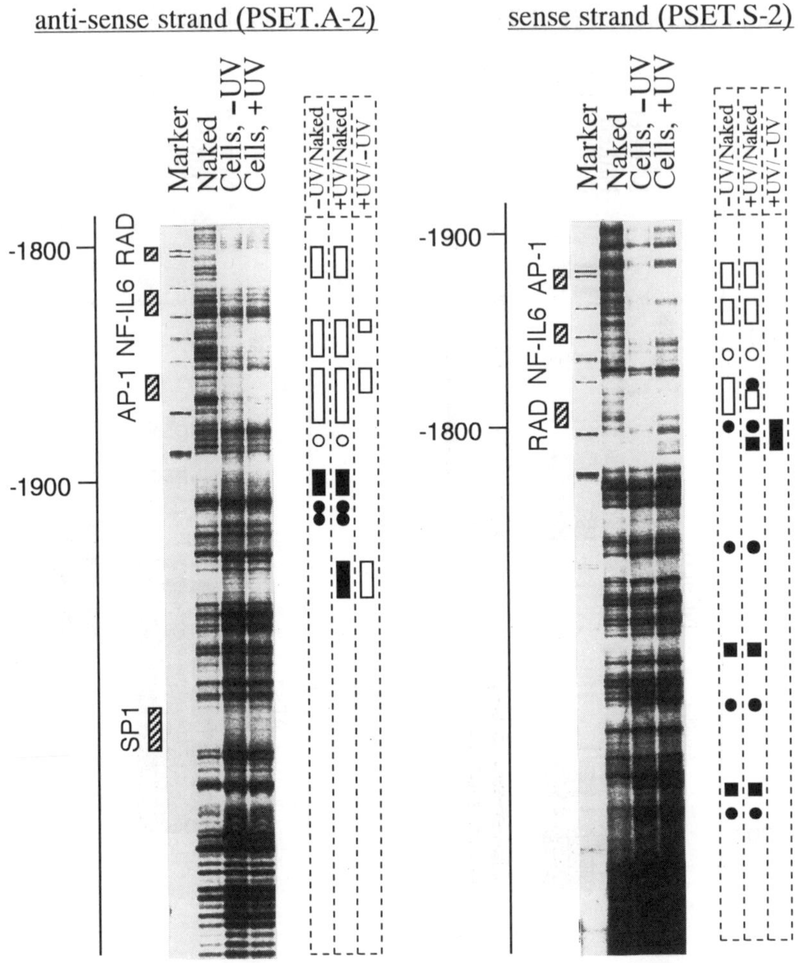

Fig.5 Genomic footprinting analysis in the region from -1950 bp to -1750 bp from the transcription initiation site.The left panel, indicated as PSET.A-2, is the autoradiogram showing the footprinting pattern on the anti-sense strand, and the right panel, indicated as PSET.S-2, is on the sense strand. Symbols and marks are ame as those in Figure 4.

To sum up, our genomic footprinting experiment could not clearly demonstrate the functional involvement of AP-1 in the UbC gene induction. However, it is still possible that the signal transduction process following UV irradiation may target the AP-1 already bound to DNA prior to gene induction. In fact, this has been shown to be the actual case for the AP-1-regulated c-*jun* gene induction.[16] In addition, we found a slight but uniform alteration of the chromatin structure in the region spanning from the AP-1 site to the NF-IL6 site after UV irradiation. AP-1 is known to crosstalk with various transcription factors, such as steroid hormone receptors, MyoD, YY1 as well as NFκB, to modulate its transactivation activity.[17] Therefore we consider the cooperative regulation of the UbC gene by AP-1 and NF-IL6 as one possibility.

Conclusions

In the present circumstances, we were able to reach the following conclusions: 1) clustered distribution of the recognition sequences for transcription factors and the presence of many SP-1 sites in the control region are necessary for UbC gene inducibility by UV light; 2) there exist recognition sequences for AP-1, NFκB and p53 in the upstream region of the UbC gene, which suggests the role of these factors in the induction of the UbC gene by UV light; 3) factors are prebound at the AP-1, NF-IL6, SP-1 and RAD sequence in the upstream region of the UbC gene; 4) a crosstalk between AP-1 and NF-IL6 is suggested.

References

1. Friedberg,E.C., Walker,G.C., and Siede,W. In: *DNA Repair and Mutagenesis*. ASM Press, Washington D.C., 1995, 595-632.
2. Sachsenmaier,C., Radler-Pohl,A., Zinck,R., Nordheim,A., Herrlich,A., and Rahmsdorf,H.J., *Cell*, 1994, **78**: 963-972.
3. Devary,Y., Gottlieb,R.A., Smeal,T., and Karin,M., *Cell*, 1992, **71**: 1081-1091.
4. Hibi,M., Lin,A., Smeal,T., Minden,A., and Karin,M., *Genes & Dev.*, 1993, 7: 2135-2148.
5. Orian,A., Whiteside,S., Israel,A., Stancovski,I. Schwartz,A.L., and Ciechanover,A., *J. Biol. Chem.*, 1995, **270**: 21707-21714.
6. Xanthoudakis,S., and Curran,T., *EMBO J.*, 1992, **11**: 653-665.
7. Kastan,M.B., Zhan,Q., El-Deiry,W.S., Carrier,F., Jacks.T., Walsh,W.V., Plunkett,B.S., Vogelstein,B., and Fornace,A.J.,Jr., *Cell*, 1992, **71**: 587- 597.
8. Ciechanover,A., and Schwartz,A.L., *FASEB J.*, 1994, **8**: 182-191.
9. Rolfe,M., Beerromero,P., Glass,S., Eckstein,J., Berdo,I., Theodoras,A., Pagano,M., and Draetta,G., *Proc. Natl. Acad. Sci. USA.*, 1995, **92**: 3264- 3268.
10. Ciechanover,A., DiGiuseppe,J.A., Bercovich,B., Orian,A., Richter,J.D., Schwartz,A.L., and Brodeur,G.M., *Proc. Natl. Acad. Sci. USA.*, 1991, **88**: 139-143.
11. Chen,Z.J., Parent,L., and Maniatis,T., *Cell*, 1996, **84**: 853-862.
12. Nenoi,M., *Int. J. Radiat. Biol.*, 1992, **61**: 205-211.
13. Nenoi,M., Cartwright,I.L., Mita,K., and Ichimura,S., *Gene*, 1996, **179**:297-299.
14. Su,W., Jackson,S., Tjian,R., and Echols,H., *Genes & Dev.*, 1991, **5**: 820-826.
15. Nenoi,M., Mita,K., Ichimura,S., and Cartwright,I.L., *Biochim. Biophys. Acta*, 1994, **1204**: 271-278.
16. Rozek,D., and Pfeifer,G.P., *Mol. Cell. Biol.*, 1993, **13**: 5490-5499.
17. Angel,P. In: *Inducible Gene Expression*, (ed. P.A.Baeuerle), Birkhauser, Boston, 1995, pp.62-92.

6. Mechanisms of Immediate-early Phase of X-ray-induced Expression of IL-1β Gene in Normal Myeloid Cells

Hiroshi ISHIHARA, Izumi TANAKA and Kazuko TSUNEOKA

The First Research Group,
National Institute of Radiological Sciences,
Anagawa 4-9-1, Inage-ku, Chiba-shi, Chiba 263, JAPAN

1. Introduction

During evolution, all bioorganisms acquire defense mechanisms against various stimuli from the environment. One form of the environmental stress, ionizing radiation, generates free radicals and results in the appearance of superoxide compounds in the presence of oxygen. Since mechanisms to control the superoxide compounds are created during the construction of aerobic respiration systems, organisms have protective mechanisms against ionizing radiation. In mammalian cells, induction of genes is observed over several hours after X-irradiation[1, 2]. These genes, called "radiation response genes", code a wide variety of proteins such as transcriptional regulators, cytokines and enzymes. Genes encoding transcriptional regulators such as NF-κB and FOS/JUN family are called "early response genes", since they are transiently activated immediately after irradiation with X-rays[3, 4]. Several hours or days after irradiation, X-irradiation induces continuous expression of genes encoding cytokines such as interleukin(IL)-1β[5] and tumor necrosis factor (TNF)-α[6]. They are occasionally called "late response genes" in contradistinction with the early response genes[7]. Since TNF-α gene are controlled by the FOS/JUN product of AP-1, it is presumed that the early response gene products regulate the late response genes[7]. However, the details of the radiation-induced mechanisms are not clear since all the radiation response genes are regulated by multiple biochemical pathways and are induced with or without radiation stresses. To analyze stress-inducible mechanisms, limitation of the genes and the pathway is necessary.

IL-1β gene is a model case for analyzing biodefense mechanism stimulated by radiation.

IL-1 is one of the cytokines secreted from differentiated macrophages and stimulates immune systems and inflammation[8]. IL-1 is known as a radioprotector, since intravenous injection of IL-1 saves lethally X-irradiated mice[9]. The fact that genes for IL-1β are induced after irradiation by X-rays[5] and ultraviolet light[10] suggests the presence of defense mechanisms against radiation stress. The induction of the genes is controlled by multiple stimulating pathways. We previously found immediate-early and later induction of the gene in spleen cells of normal mouse after X-irradiation[5]. The molecular mechanisms are analyzed to investigate radiation-induced response.

2. Later Response of the IL-1β Gene

Several days after whole-body X-irradiation of normal C3H/He mice at a dose of 4-8 Gy, temporal accumulation of IL-1β mRNA in spleen cells was observed[5]. Similar remarkable accumulation of messages for other cytokines including IL-1α was not observed. To estimate the cause of the temporal accumulation, we investigated the change in spleen cells after whole-body irradiation of C3H mice at a sublethal dose of 3Gy. When the spleen cells were analyzed by *in situ* hybridizaton probed by antisense IL-1β RNA, remarkable accumulation of IL-1β message was observed in a limited number of spleen cells from mice irradiated after 8 days (Fig. 1). The cells with a high degree of expression show similar morphology to monocyte/macrophage. In contrast, such accumulation was not observed in any of the cells from non-irradiated mice or irradiated mice at Day 8. Additionally, the monocyte/macrophage rate in spleen increased after irradiation and peaked at 8 days[11]. These results suggest that the whole-body irradiation systemically stimulates differentiation of spleen macrophages which express large amounts of IL-1β mRNA.

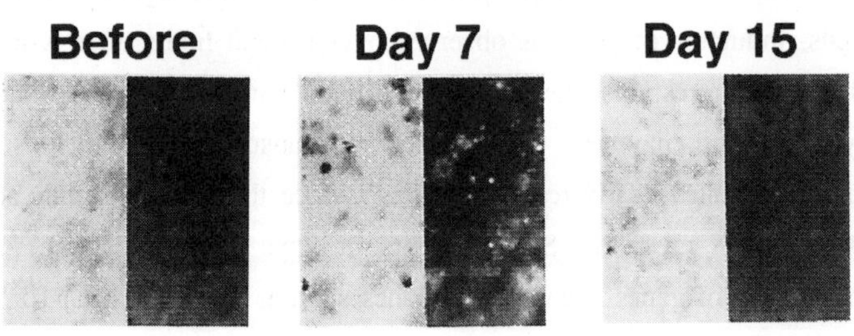

Fig. 1 Expression of IL-1β mRNA in heterogeneous spleen cells. Spleen cells before and after indicated days of whole-body radiation were used for *in situ* hybridization analysis. Each pair of photomicrographs consists of light-field (left) and dark-field (right) views of microradiographs of the same area.

3. Immediate-early Induction of IL-1β Gene

(a) The immediate-early response occurs in monocyte-macrophages

In contrast to the fact that the late response was the result of several systemic regulation mechanisms, the immediate-early accumulation of IL-1β seemed to be controlled by a limited number of biochemical pathways. The immediate-early accumulation was first observed when the primary cultured spleen cells were irradiated at a high dose at 20 Gy of X-rays[5]. Similar accumulation in the cultured spleen cells was detected 30 minutes after irradiation of a lower dose at 3 Gy of X-rays[12]. When normal C3H/He mice were whole-body-irradiated at 18.5 Gy, accumulation of IL-1β mRNA was observed in spleen cells[12].

Spleen contains various type of cells including lymphoid and myeloid cells at diverged differentiation stages. To determine the type of cells which cause immediate-early accumulation of IL-1β message, we separated irradiated and non-irradiated cells from normal spleen using surface-specific antibody and magnetic beads. Monocytic, T-lymphoid and B-lymphoid cells were separately fixed onto slide glass and analyzed by *in situ* hybridization probed with antisense IL-1β RNA. As shown in Fig. 2, remarkable accumulation of IL-1β was observed in irradiated monocyte-macrophages, whereas the non-irradiated monocyte-macrophages express low level of the

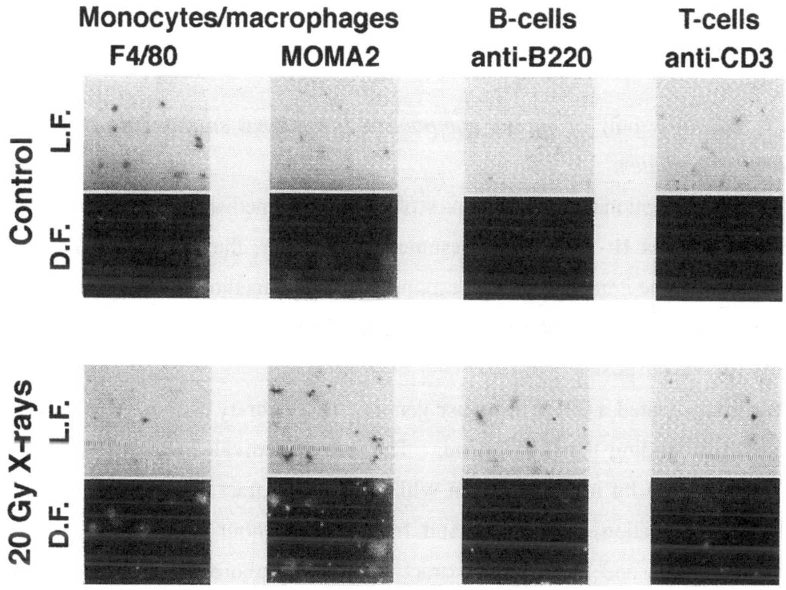

Fig. 2 Detection of IL-1β mRNA in fractionated spleen cells. Normal spleen cells before and after irradiation were separated using the indicated antibodies and analyzed by *in situ* hybridization.

message. In contrast, the message was not detected in lymphoid cells. This shows that the immediate-early accumulation of IL-1β gene observed in normal spleen cells is due to monocyte-macrophage. Similar immediate-early response in IL-1β mRNA was also observed in established macrophage-like leukemia cells , L-8704[13].

(b) The immediate-early response is regulated at the transcriptional and post-transcriptional level

It is known that temporal induction of the IL-1β gene is transcriptionally regulated by the binding of nuclear proteins at the upstream region in mouse[14] and human[15-17]. Additionally, the cellular level of the message was also regulated post-transcriptionally using unique AUUUA motif at the 3' untranslated region of the message[18]. To clarify the contribution of the above to the immediate-early response, the following quantitative assays were performed.

The nuclear run-on assay was performed using isolated nuclei from non-irradiated or irradiated L-8704 cells (Fig. 3A). Increase in levels of IL-1β transcript was observed by using nuclei prepared from cells immediately after irradiation, indicating that the regulation at the transcriptional level contributed to the immediate-early response.

To estimate the contribution of post-transcriptional regulation to the immediate-early accumulation of IL-1 β mRNA, quantitative Northern analysis was done (Fig. 3B). After IL-1β mRNA reaches maximal level by treatment with dbcAMP in L-8704 cells, decrease rate of the message was measured. Decrease in the message was prevented by irradiation, showing that post-transcriptional regulation also contributes to the immediate-early response.

(c) Specific binding activity of nuclear components to the areas surrounding IL-1β gene decreased immediately after irradiation

Since transcriptional regulation takes place in the immediate-early response, the presence of a functional area besides IL-1β gene is presumed. However, the convenient reporter assay is not suitable for analyzing the immediate-early response, since the method measures accumulated changes over several days incubation of the cells following irradiation. Therefore, we used nuclear extract-binding assay for our purpose.

We first prepared a C3H/He mouse genomic DNA library with λEMBL3 vector and cloned the 30 kb area surrounding the IL-1β gene. The area was divided and fragmented by restriction enzymes up to 200-400 bp in size. Mean while, nuclear extract was prepared from L-8704 cells immediately after irradiation by X-rays and from cells without irradiation. Mixtures of radio-labeled DNA fragments and the nuclear extract were electrophoresed, and the effects of different pretreatments on the nuclear extract analyzed. Since the DNA-nuclear extract complex makes multiple bands of various mobility, we measured the amount of free DNA which was not absorbed to the binding. The comparisons were successful, the result of the application of small amounts of

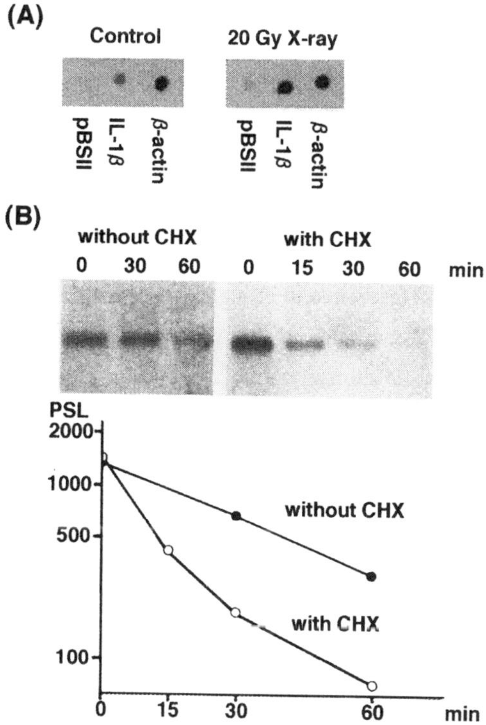

Fig. 3 Analysis of regulation mechanisms of IL-1β mRNA accumulation after X-irradiation.
(A) Incorporation of radioactive phosphate into preRNA was performed using Isolated nuclei prepared from non-irradiated or irradiated L8704 cells. As probes, the products were used for hybridization of the indicated DNAs fixed on the nitrocellulose filter. (B) Effect of X-irradiation on the rate of decrease in IL-1β mRNA level. After cellular level of IL-1β mRNA reached maximum by dbcAMP treatment, the cells were irradiated with or without cycloheximide (CHX). Quantitative Northern analysis was performed using RNAs prepared at the indicated min following radiation.

labeled DNA (0.01p mole) and high concentration of agarose gel.

We found three unique fragments besides IL-1β gene binding to the nuclear extracts and determined the physical map of the fragments (Fig. 4). The nucleotide sequence of these three fragments showed no similarity with known functional motifs. When nuclear extract from the non-irradiated cells was used, components of the nuclear extract associated with these fragments. The binding disappeared in the presence of an excess amount of non-labeled DNA at the same sequence as a competitor, showing that the binding is sequence-specific. The pretreatment of the nuclear extract with proteinase K eliminated the specific binding, indicating that the binding was due to the protein included in the nuclear. In contrast, nuclear extract from the irradiated cells did not show specific binding activity. These results suggest that some nuclear protein in non-irradiated cells disappear immediately after irradiation by X-ray in the macrophage-like L-8704 cells.

4. Conclusion and Discussion

We previously found temporal accumulation of IL-1β mRNA in spleen several days after whole-body irradiation of x-ray[5]. This is due to change in the number and differentiation stage of spleen macrophages[11]. We also found that IL-1β mRNA accumulated immediately after X-irradiation in cultured spleen cells from normal mouse[5]. It was proved that the immediate-early phase of induction occurred in monocyte-macrophage and was regulated at the transcriptional and post-transcriptional level[12]. Immediately after irradiation by X-rays, disappearance of nuclear protein(s) which bind specifically to the upstream regions of IL-1β gene was observed. The nucleotide sequence of the binding target fragment show no similarity with previously known sequences.

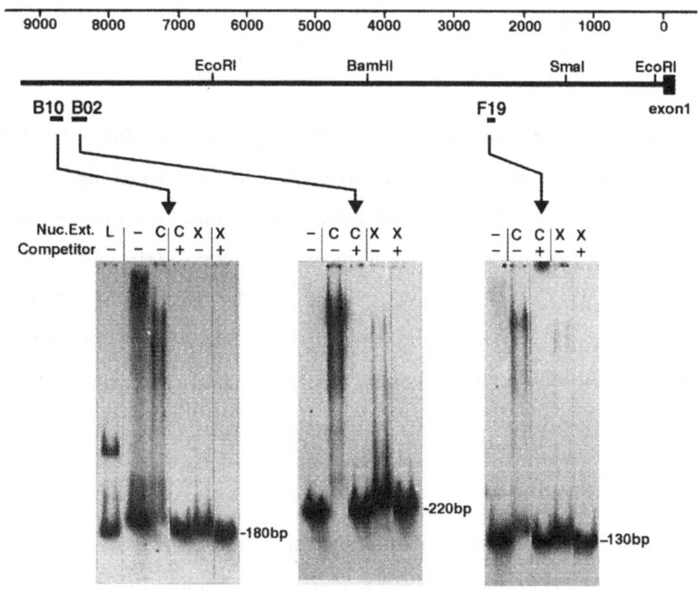

Fig. 4 Decrease in the binding acitivity to the IL-1β gene locus immediately after X-irradiation. Specific binding activity in nuclear extract to the DNA fragments corresponding to IL-1β gene locus were analyzed by EMSA assay. Non-labeled fragments were used as binding competitors, and specific binding sites were mapped on the IL-1β gene locus.

The immediate-early induction of the IL-1β gene occurs within 30 min after exposure to X-rays. At this time, some of the fos/jun family oncogenes which encode transcriptional regulators are simultaneously activated[2, 3, 7]. It is believed that the sequential induction of various genes are triggered by the activation of the oncogenes. Additionally, Fos/Jun protein can activate fos/jun gene[19], and exogenous the IL-1β can stimulate expression of IL-1β gene[20]. Therefore, the immediate-early induction of the IL-1β gene is probably the beginning of a subsequent multiple cascade reaction (Fig. 5). Interestingly, IL-1 induces a gene for superoxide dismutase that can eliminate superoxides, which are known to be a major toxic product of ionizing radiation[21]. These findings support the possibility of the presence of protective mechanisms against ionizing radiation by the removal of superoxides *via* activation of the IL-1β gene.

Although the late phase induction of the IL-1β gene occurs after increase in the specific bindings of nuclear proteins such as NF-IL6[15], decrease in binding is observed immediately after irradiation. The target areas of the early phase have no consensus sequence for binding to NF-κB nor any other functional proteins, whereas binding activity to NF-κB is observed immediately after irradiation[22]. This suggests that the nuclear protein(s) and/or the target sequences for radiation-inducible disappearance of binding are controlled by unknown molecular mechanisms. The undetermined protein is the end point of the primary signal transduction pathway immediately after irradiation by X-rays. Further investigation focusing on immediate-early induction will be useful to clarify the primary induction mechanism against X-rays, to analyze mechanisms such as radio-protection and -sensitization, and to design a novel radioprotector.

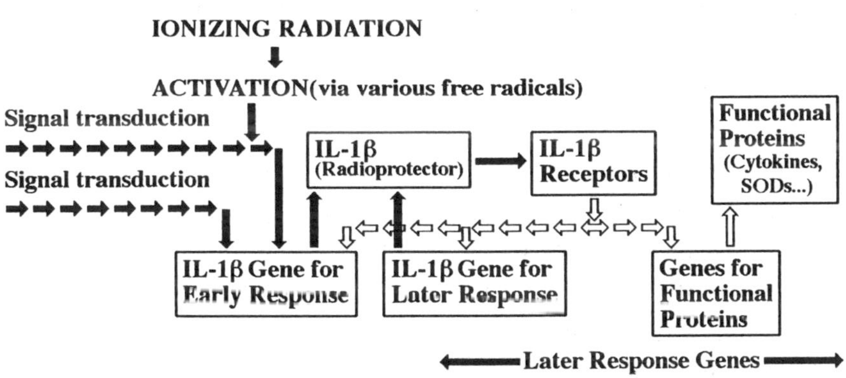

Fig. 5 Scheme of induction cascade of IL-1β gene. Ionizing radiation activates some signal transduction pathways. IL-1β gene is controlled by immediate-early and late phases of induction. IL-1β stimulates various induction pathways and can survive damage by X-irradiation.

References

1. Fornace Jr, A. J. (1992) *Annu. Rev. Genet.,* **26**: 507-526.

2. Herrlich, P., Ponta, H., and Rahmsdorf, H. J. (1992) *Rev. Physiol. Biochem. Pharmacol.,* **119**: 187-223.

3. Sherman, M. L., Datta, R., Hallahan, D. E., Weichselbaum, R. R., and Kufe, D. W. (1990) *Proc. Natl. Acad. Sci. USA.,* **87**: 5663-5666.

4. Brach, M. A., Hass, R., Sherman, M. L., Gunji, H., Weichselbaum, R., and Kufe, D. (1991) *J. Clin. Invest.,* **88**: 691-695.

5. Ishihara, H., Tsuneoka, K., Dimchev, A. B., and Shikita, M. (1993) *Radiat. Res.,* **133**: 321-326.

6. Hallahan, D. E., Spriggs, D. R., Beckett, M. A., and Kufe, D. W. (1989) *Proc. Natl. Acad. Sci. USA.,* **86**: 1014-1017.

7. Weichselbaum, R. R., Hállahan, D. E., Sukhatme, V., Dritschilo, A., Sherman, M. L., and Kufe, W. (1991) *J. Natl. Cancer Inst.,* **83**: 480-484.

8. Bagby Jr, G. C. (1989) *Blood Rev.,* **3**: 152-161.

9. Neta, R., Douches, S., and Oppenheim, J. J. (1986) *J. Immunol.,* **136**: 2483-2485.

10. Kupper, T. S., Chua, A. O., Flood, P., McGuire, J., and Gubler, U. (1987) *J. Clin. Invest.,* **80**: 430-436.

11. Nemoto, K., Ishihara, H., Tanaka, I., Suzuki, G., Tsuneoka, K., Yoshida, K., and Ohtsu. H. (1995) *J. Radiat. Res.,* **36**: 125-133.

12. Ishihara, H., Tanaka, I., Nemoto, K., Tsuneoka, K., Cheeramakara, C., Yoshida, K., and Ohtsu, H. (1995) *J. Radiat. Res.,* **36**: 112-124.

13. Ishihara, H., Yoshida, K., Nemoto, K., Tsuneoka, K., and Shikita, M. (1993) *Radiat. Res.,* **135**: 394-399.

14. Godambe,S.A., Chaplin,D.D., Takava,T., and Bellone,C.J. (1994) *J. Immunul.,* **153**: 143-152.

15. Shirakawa, F., Saito, K., Bonagura, C. A., Galson, D. L., Fenton, M. J., Webb, A. C., and Auron, P. E. (1993) *Mol. Cell. Biol.,* **13**: 1332-1344.

16. Cogswell, J. P., Godlevski, M. M., Wisely, G. B., Clay, W. C., Leesnitzer, L. M., Ways, J. P., and Gray, J. G. (1994) *J. Immunol.,* **153**: 712-723.

17. Zhang,Y., and Rom,W.N. (1993) *Mol. Cell. Biol.,* **13**: 3831-3837.

18. Capt, D., Beutler, B., Hartog, K., Thayer, R., Shimer, S. B., and Cerami, A. (1986) *Proc. Natl. Acad. Sci. USA.,* **83**: 1670-1674.

19. Shah, G., Ghosho, R., Amstad, P. A., and Cerutti, P. A. (1993) *Cancer Res.,* **53**: 38-45.

20. Fenton, M. J., Vermeulen, M. W., Clark, B. D., Webb, A. C., and Auron, P. E. (1988) *J. Immunol.,* **140**: 2267-2273.

21. Masuda, A., Longo, D. L., Kobayashi, Y., Appela, E., Oppenheim, J. J., and Matsushima, K. (1988) *FASEB J.,* **2**: 3087-3091.

III

Apoptosis as a Biodefence Mechanism

7. An Overview of Apoptosis with Emphasis on Its Biodefence Functions

John KERR

Formerly Department of Pathology, University of Queensland Medical School, Herston 4006, Australia

In synopsis, apoptosis is a distinct type of cellular death that occurs in normal tissues as well as under certain pathological conditions (1-3). Its features suggest active cellular self-destruction, not degeneration as seen in classical necrosis (1,2). In many situations, it plays an opposite role to mitosis in regulating tissue size (l,2).

Where apoptosis occurs physiologically, it clearly serves a homoeostatic function. However, in at least some of the circumstances in which it occurs pathologically it can also be regarded as a homoeostatic phenomenon in that it brings about the deletion of cells whose survival might endanger the host animal (4). That is, in these cases, it constitutes a biodefence mechanism. For example, target cell destruction induced by cytotoxic T lymphocytes and the deletion of cells that follows administration of small to moderate doses of radiation are both mediated by apoptosis (4). The former is an essential component of the body's defences against viral infections, whereas the latter helps prevent the persistence of potentially dangerous cells with unrepaired damage to DNA.

Features of apoptosis

Apoptosis was originally defined and distinguished from necrosis on the basis of its ultrastructure (1), and electron microscopy remains a reliable method for identifying the process (5). It characteristically affects scattered individual cells. The earliest recognisable morphological change comprises the presence of sharply delineated, uniformly finely granular masses of compacted chromatin that lie against the inner surface of the nuclear envelope. The cytoplasm then condenses, the nucleus shows convolution of its outline followed by budding to give rise to fragments that are still surrounded by double membranes, and protuberances of varying size develop on the surface of the cell. The latter separate with plasmalemmal sealing to produce membrane-bounded apoptotic bodies in which the organelles, including lysosomes, are still intact.

In tissues, these apoptotic bodies are quickly phagocytosed and digested by nearby cells, a variety of cell types, including epithelial cells, being implicated. There is no accompanying inflammation. The viable cells close ranks and the cell deletion is achieved without disruption of overall tissue architecture. The whole process is rapid, the cellular condensation and budding that give rise to apoptotic bodies being completed within minutes and the digestion of phagocytosed bodies taking only a few hours (2). The small size of many apoptotic bodies, the speed with which they are removed and the absence of inflammation make apoptosis inconspicuous histologically. This needs to be taken into account when assessing the kinetic significance of the presence of apoptotic bodies in a particular situation.

Necrosis, in contrast, tends to affect groups of adjoining cells in a tissue and is usually associated with the occurrence of inflammation. Disturbance of membrane function with cellular swelling, followed by membrane disruption, lysosomal enzyme release and cellular degradation are characteristic features (2,5).

The biochemical processes that underlie the sequence of morphological events occurring in apoptosis are incompletely understood. The most extensively studied involves the nuclear DNA. There is double-strand cleavage at the linker regions between nucleosomes leading to the production of oligonucleosome-sized fragments (6-9). These are conveniently demonstrated by agarose gel electrophoresis, where a so-called ladder develops. In necrosis, the DNA is degraded to a continuous spectrum of fragment sizes, a diffuse smear being observed on electrophoresis (4). There is still debate about the identity of the endonuclease involved in the internucleosomal cleavage in apoptosis (9). Further, it has been shown that DNA is also cleaved into large, 50-300 kilobase-pair lengths and it has been questioned whether internucleosomal cleavage is invariable (9).

The nuclear lamina disappears in apoptosis, as a result of proteolytic cleavage of lamins, and it has been suggested that this may play a part in producing the nuclear changes observed morphologically (10). The enzyme responsible, Mch-2 (11), is a member of the ICE family of proteases, which will be briefly discussed later. Little appears to be known about the mechanism of the cytoplasmic condensation.

The violent protrusion of the cell surface that precedes apoptotic body formation obviously indicates involvement of the cytoskeleton. ß tubulin messenger RNA increases early and this is followed by increased ß tubulin synthesis (12). Inhibitors of actin polymerization such as cytochalasin B prevent the cellular budding without interfering with the DNA cleavage (13). The fodrin meshwork, which lies under the plasma membrane and which is concerned with the maintenance of cell shape, is enzymatically degraded (14).

It has been stressed that apoptotic bodies formed in tissues are rapidly phagocytosed. When phagocytosis fails to take place, which is often the case in cell cultures, the bodies spontaneously degenerate within a few hours (2,5). If this were to happen *in vivo*, inflammation would ensue. Several mechanisms for recognition of apoptotic bodies by phagocytic cells appear to have evolved in mammals (15). The most extensively studied involves translocation, during apoptosis, of phosphatidylserine from the inner to the outer leaflet of the lipid bilayer of the plasma membrane. This change is recognised by receptors on the phagocytic cell.

Physiological occurrence of apoptosis

Apoptosis is responsible for deletion of cells in normal animals, where it serves a variety of physiological functions (2,4,16,17). An important message to emerge from consideration of its physiological occurrence is that it is a regulated phenomenon. Some of the most dramatic examples are seen during development. For example, the rapid regression of the tadpole tail that takes place at metamorphosis is mediated by apoptosis, and apoptosis is responsible for morphogenetic processes such as deletion of interdigital webs during mammalian development.

In adult mammals, apoptosis occurs continually in both slowly and rapidly proliferating tissues. In the former, which comprise most of the tissues of the body, it is balanced by mitosis under steady state conditions. However, in tissues of the latter type, such as bone marrow and intestinal crypts, much of the mitosis is related to migration of cells from the tissue. The extent of apoptosis in adult tissues is regulated both by endocrine and paracrine factors (16) and by intrinsic cellular proteins, which modulate the effect of addition or withdrawal of the factors. Paramount amongst these intrinsic proteins are members of the Bcl-2 family (18). The *bcl-2* gene was proposed as a proto-oncogene because of its location at the breakpoint in the 14;18 chromosome translocation that occurs in human follicular B cell lymphoma. However, its expression was found to promote cell survival following grown factor withdrawal rather than the expected cell proliferation (19). It was subsequently shown that Bcl-2 specifically inhibits apoptosis (20), and a search for proteins showing homology with Bcl-2 has disclosed a family of apoptosis regulatory proteins (18). Some, such as Bcl-XL, inhibit apoptosis following a stimulus, whereas others, such as Bax, facilitate its occurrence. Differential expression of proteins that modulate the apoptotic response in the individual members of a cell population can lead to the selective occurrence of apoptosis in a tissue.

Apoptosis is responsible for various normal involutional processes in adult mammals such as regression of the lactating breast after weaning and atresia of ovarian follicles. It serves a variety of regulatory roles in the immune system, including the deletion of potentially autoreactive B and T lymphocytes. Lastly, it accounts for elimination of effete cells such as ageing neutrophil leukocytes and megakaryocytes that have released their platelets.

Biodefence functions of apoptosis

There is abundant evidence that target cell death induced by cytotoxic T lymphocytes takes the form of apoptosis (21). The main function of the cellular immune system under natural conditions is the elimination of virus-infected cells, thus preventing the continued production of infectious viral particles. It is attractive to speculate that this system made opportunistic use of the apoptotic pathway during evolution to enable it to eliminate such cells with a minimum of tissue disruption. There is recent molecular evidence to support this hypothesis.

Two mechanisms of activation of apoptosis by cytotoxic T lymphocytes have been identified (22). Both are initiated by recognition of antigenic peptide presented by MHC on the surface of the target cell by the T cell receptor on the effector cell. In the first mechanism, granzyme B and perforin are released by the lymphocyte. The latter allows entry of the target cell by the former, which has been shown to cleave its substrates after aspartic acid residues. This property is shared by members of the ICE family of proteases (23), which are mammalian homologues of CED-3, a protease required for the occurrence of cell death during normal development of the nematode worm *Caenorhabditis elegans*. A number of ICE-family proteases have been shown to be involved in the execution of apoptosis (23). They are produced as inactive pro-enzymes and either activate themselves or are activated by other ICE-family members. Granzyme B cleaves and activates at least five apoptosis-related enzymes in the ICE family (23,24). The second mechanism of induction of apoptosis by cytotoxic T lymphocytes involves the Fas or APO-1 receptor (22). This belongs to the tumour necrosis factor receptor/nerve growth factor receptor superfamily of cell surface proteins. Upon contact with its target, the effector lymphocyte rapidly increases its synthesis of Fas ligand and expresses it on its surface. Ligation of the Fas receptor on the target then triggers apoptosis. The cytoplasmic death domain of the Fas receptor shows homology with reaper, a protein involved in the induction of apoptosis in *Drosophila* (25). It has been suggested that a reaper-like cell death effector was linked during mammalian evolution to a surface receptor allowing it to be activated by cell surface signals (25). Thus both mechanisms of T cell killing may exploit primitive, phylogenetically conserved methods of activating apoptosis.

Now we turn to the induction of apoptosis by radiation. Whilst very large doses of radiation cause necrosis, small to moderate doses trigger apoptosis (2,8,26-28). Only a restricted number of cell types is susceptible (26,27). These include certain members of rapidly proliferating cell populations such as intestinal crypts and seminiferous tubules, cells in many parts of the embryo and fetus, and lymphoid cells. It is noteworthy that the latter die in interphase. Several other types of cells can be induced to undergo apoptosis by radiation, albeit when given in rather larger doses than those required to induce apoptosis of the cell types listed above; their consideration will be deferred until later.

In the proliferating cell populations, susceptibility is not simply linked to the proliferation, which might be expected to favour the fixation of DNA lesions. For example, following exposure of the small intestine to radiation, extensive apoptosis is seen near the base of the crypts, but the process is present to only a much more limited extent up their sides, despite the active mitosis that normally occurs in the latter situation (28,29). It can be argued that the propensity for a cell to undergo apoptosis after irradiation is a measure of the danger of its persisting in mutant form (26). Survival of cells with unrepaired DNA damage near the base of the small intestinal crypt, where the stem cells reside, would lead to immortalization of the lesion and possibly to the later development of cancer (29). Mutant cells up the sides of the crypt, on the other hand, will soon move onto the villus and be shed into the intestinal lumen. It is of interest that, in the large intestinal crypt, radiation-induced apoptosis mostly occurs above the stem cell region, the stem cells being protected by high levels of Bcl-2 (29). During evolution, it may have been necessary for colonic stem cells to develop protective mechanisms against apoptosis-inducing toxins produced by luminal bacteria (29). But there is a price to pay for this protection; colonic adenocarcinoma is very common whilst small intestinal adenocarcinoma is rare (28). In the seminiferous tubule, radiation induces apoptosis of spermatogonia and of early preleptotene primary spermatocytes, the latter being the cells that undertake the final DNA synthesis in the germ cell lineage (27). The danger of persistence of these cells if they carry DNA lesions is obvious; they would give rise to mutant gametes. Persisting mutant cells in the embryo would produce large numbers of mutant cells in the adult. And lymphocytes are unique; mutations leading to defective self-recognition might result in autoimmunity if the affected cells were not eliminated (30).

How does DNA damage trigger apoptosis? There is strong evidence that the product of the p53 tumour suppressor gene is involved (31,32). In the irradiated murine small intestine, immunohistochemistry reveals elevated levels of p53 protein in cells near the base of the crypts, the spatial distribution of the p53 staining almost exactly matching the distribution of apoptosis (28,29). In p53 deficient mice, the small intestinal crypt cells are resistant to the induction of apoptosis by radiation (29,33), whereas levels of spontaneous apoptosis in the crypts of these animals are normal (28). Likewise, thymocytes from animals lacking wild-type p53 are resistant to the induction of apoptosis by radiation, but show a normal susceptibility to its induction by glucocorticoids (34,35). Lane (31) has proposed that DNA damage leads to an increase in p53 protein through a post-translational stabilisation mechanism and this arrests the cell cycle in G1, allowing extra time for DNA repair. If repair fails, p53 may then trigger deletion of the cell by apoptosis. If this proposal is correct, p53 abnormalities would be expected to allow cells with DNA damage to persist, thus favouring tumour development and progression. The latter appears to be the case (36-39).

It was noted earlier, when the cell types susceptible to the induction of apoptosis by very small doses of radiation were being enumerated, that several other types of cells can be induced to undergo apoptosis by irradiation, though larger doses are required. These include the differentiated acinar cells of salivary (40) and lacrimal glands (41) and endothelial cells (42). The biological reason for the susceptibility of these cells is not clear. Fuks (43, and see paper this volume) has shown that the induction of apoptosis of endothelial cells by radiation involves the sphingomyelin-ceramide pathway rather than p53.

Conclusion

The essence of apoptosis is that it is an active, regulated phenomenon. That is why it is an interesting object of study. Preoccupation with the classical paradigm of cell death, which viewed it as the inevitable outcome of massive disruption of the biochemical machinery of the cell by severe injury, inhibited studies of cell death for many years.

References

1. Kerr, J.F.R., Wyllie, A.H., Currie, A.R. Br. J. Cancer. 1972, 26: 239-57.
2. Wyllie, A.H., Kerr, J.F.R., Currie, A.R. Int.Rev.Cytol. 1980, 68: 251-306.
3. Thompson, C.B. Science. 1995, 267: 1456-62.
4. Kerr, J.F.R., Harmon, B.V. In: Apoptosis: The Molecular Basis of Cell Death. Tomei, L.D., Cope, F.O. (Eds.), Cold Spring Harbor LaboratoryPress, New York, 1991, 5-29.
5. Kerr, J.F.R., Gobé G.C., Winterford, C.M., Harmon, B.V. In: Methods in Cell Biology, Vol.46. Schwartz, L.M., Osborne, B,A. (Eds.), Academic Press, San Diego, 1995, 1-27.
6. Wyllie, A.H. Nature. 1980, 284: 555-6.
7. Umansky, S.R. J. Theor. Biol. 1982, 97: 591-602.
8. Yamada, T., Ohyama, H. Int. J. Radiat. Biol. 1988, 53: 65-75.
9. Bortner, C.D., Oldenburg, N.B.E., Cidlowski, J.A. Trends Cell Biol. 1995, 5: 21-6.
10. Lazebnik, Y.A., Cole, S., Cooke, C.A., Nelson, W.G., Earnshaw, W.C. J. Cell Biol. 1993, 123: 7-22.
11. Takahashi, A., Earnshaw, W.C. Curr. Opin. Genet. Dev. 1996, 6: 50-5.
12. Pittman, S.M., Geyp, M., Tynan, S.J., Gramacho, C.M., Strickland, D.H. et al. In: Programmed Cell Death: The Cellular and Molecular Biology of Apoptosis. Lavin, M., Watters, D. (Eds.), Harwood Academic Publishers, Switzerland, 1993, 315-23.
13. Cotter, T.G., Lennon, S.V., Glynn, J.M., Green, D.R. Cancer Res. 1992, 52: 997-1005.
14. Martin, S.J., O'Brien, G.A., Nishioka, W.K., McGahon A.J., Mahboubi, A. et al. J. Biol. Chem. 1995, 270: 6425-8.
15. Savill, J., Fadok, V., Henson, P., Haslett, C. Immunol. Today. 1993, 14: 131-6.
16. Raff, M.C. Nature. 1992, 356: 397-400.

17. Owen, J.J.T., Allan, I.M., Moore, N.C., Jenkinson, E.J. In: Programmed Cell Death: The Cellular and Molecular Biology of Apoptosis. Lavin, M., Watters, D. (Eds.), Harwood Academic Publishers, Switzerland, 1993, 219-27.
18. Nuñez, G., Clarke, M.F. Trends Cell Biol. 1994, 4: 399-403.
19. Vaux, D.L., Cory, S., Adams, J.M. Nature. 1988, 335: 440-2.
20. Hockenbery, D., Nuñez, G., Milliman, C., Schreiber, R.D., Korsmeyer, S.J. Nature. 1990, 348: 334-6.
21. Duke, R.C. In: Apoptosis: The Molecular Basis of Cell Death. Tomei, L.D., Cope, F.O. (Eds.), Cold Spring Harbor Laboratory Press, New York, 1991, 209-26.
22. Nagata, S., Golstein, P. Science. 1995, 267: 1449-56.
23. Kumar, S., Lavin, M.F. Cell Death Differ. 1996, 3: 255-67.
24. Darmon, A.J., Nicholson, D.W., Bleackley, R.C. Nature. 1995, 377: 446-8
25. Golstein, P., Marguet, D., Depraetere, V. Cell. 1995, 81: 185-6.
26. Kerr, J.F.R., Winterford, C.M., Harmon, B.V. Cancer. 1994, 73: 2013- 26.
27. Allan, D.J. Int.J.Radiat.Biol. 1992, 62: 145-52.
28. Potten, C.S., Merritt, A., Hickman, J., Hall, P., Faranda, A. Int.J. Radiat. Biol. 1994, 65: 71-8.
29. Hickman, J.A., Potten, C.S., Merritt, A.J., Fisher, T.C. Phil. Trans. R. Soc. Lond. B. 1994, 345: 319-25.
30. Cohen, J.J., Duke, R.C. Annu.Rev.Immunol. 1992, 10: 267-93.
31. Lane, D.P. Nature. 1992, 358: 15-6.
32. Lane, D.P. Nature. 1993, 362: 786-7.
33 Clarke, A.R., Gledhill, S., Hooper, M.L., Bird, C.C., Wyllie, A.H. Oncogene. 1994, 9: 1767-73.
34. Clarke, A.R., Purdie, C.A., Harrison, D.J., Morris, R.G., Bird, C.C. et al. Nature. 1993, 362: 849-52.
35. Lowe, S.W., Schmitt, E.M., Smith, S.W., Osborne, B.A., Jacks, T. Nature. 1993, 362: 847-9.
36. Donehower, L.A., Harvey, M., Slagle, B.L., McArthur, M.J., Montgomery, C.A. et al. Nature. 1992, 356: 215-21.
37. Ziegler, A., Jonason, A.S., Leffell, D.J., Simon, J.A., Sharma, H.W. et al. Nature. 1994, 372: 773-6.
38. Kamb, A. Nature. 1994, 372: 730-1.
39. Symonds, H., Krall, L., Remington, L., Saenz-Robles, M., Lowe, S. et al. Cell. 1994, 78: 703-11.
40. Sodicoff, M., Pratt, N.E., Sholley, M.M. Radiat.Res. 1974, 58: 196-208.
41. Gazda, M.J., Schultheiss, T.E., Stephens, L.C.,. Ang, K.K., Peters, L.J. Int. J. Radiat. Oncol. Biol. Phys. 1992, 24: 693-7.
42. Fuks, Z., Persaud, R.S., Alfieri, A., McLoughlin, M., Ehleiter, D. et al. Cancer Res. 1994, 54: 2582 90.
43. Haimovitz Friedman, A., Kan, C.C., Ehleiter, D., Persaud, R.S., McLoughlin, M. et al. J. Exp. Med. 1994, 180: 525-35.

8. Radiation-induced Apoptosis in Rat Thymocytes

Harumi OHYAMA[1], Takashi SHIMOKAWA[1,2,3,4], Hiroko HAMA-INABA[5], Nobuo IKOTA[5], Osami YUKAWA[6] and Takeshi YAMADA[2]

[1]Div. Radiobiol. & Biodosimetry, [5]Dept. Bioregulation Res., [6]Div. Biol.& Oncol. , NIRS, [2]Toho Univ. Graduate Sch. Biomol. Sci., [3]Biochem. Div., Natl. Cancer Ctr. Res. Inst., [4]Dept. Radiol. Health, Faculty of Med. Univ. Tokyo

Thymic small lymphocytes, commonly known as thymocytes, are known to be one of the most radiosensitive cells in the body. They undergo cell death within a short time after moderate dose X-irradiation without intervention of mitosis, this is termed "interphase death" in radiation biology.

In 1980 we succeeded in separating the dead cell fraction from the irradiated thymocyte suspensions by centrifugation on Percoll gradient[1). Examination of the separated dead cells led us to notice that radiation-induced interphase death of thymocytes accompanied a sharp but transient increase in buoyant density, concomitant with the disappearance of microvilli in the cell surface[2)] and the distinctive reduction in cell volume[3)]. The chromatin DNA in the dead cells was fragmented into oligonucleosomal sizes having a molecular weight sufficiently low to resist sedimentation[4)]. These results indicate clearly that thymocyte interphase death is a typical example of apoptosis[5, 6)], a mode of the cell death, defined by Kerr, Wyllie and Currie in 1972[7)] as distinctly different from necrosis.

Although intensive studies have been conducted, the molecular mechanism underlying the radiation-induced apoptosis remains to be investigated. Recently, oxidative stress has been suggested to be a mediator of apoptotic cell death. Radiation could be considered as one form of oxidative stress, because ionizing radiation generate reactive oxygen species (ROS). It is of special interest, therefore to explore the role of oxidative stress in radiation-induced apoptosis.

Here we review our research on radiation-induced apoptosis in thymocytes and present recent results with peroxide formation and its role in apoptosis.

1. Characteristic Features of Radiation-induced Apoptosis in Thymocytes

1.1 Increase in Buoyant Density

Rat thymocytes undergo apoptosis within a few hours after *in vitro* X-irradiation. Percoll density gradient centrifugation allowed us to find that apoptotic cells had buoyant density distinctly higher than that of viable preapototic cells[1)].

1.2 Cell Size Reduction

The increase in the buoyant density could be explained by the fact that the irradiated thymocytes condensed to a definite size, in association with apoptosis[3]. Measurement of cell size distribution demonstrated that the irradiated thymocyte suspension consisted of two discrete subpopulations, one of normal size and the other of smaller size as shown in Fig. 1a. Reduction of cell size is a characteristic morphological feature of apoptosis.

The number of smaller cells increased parallel to a rise in apoptotic cell counts with incubation time *in vitro* and radiation dose. The dead cell population separated by Percoll density gradient centrifugation was found to correspond exactly to smaller sized ones (Fig. 1b).

Cell size distribution analysis after a whole-body X-irradiation revealed that thymocytes in thymus gland were fragmented into apoptotic bodies with various sizes beyond the state of cellular condensation observed under *in vitro* conditions [8, 9].

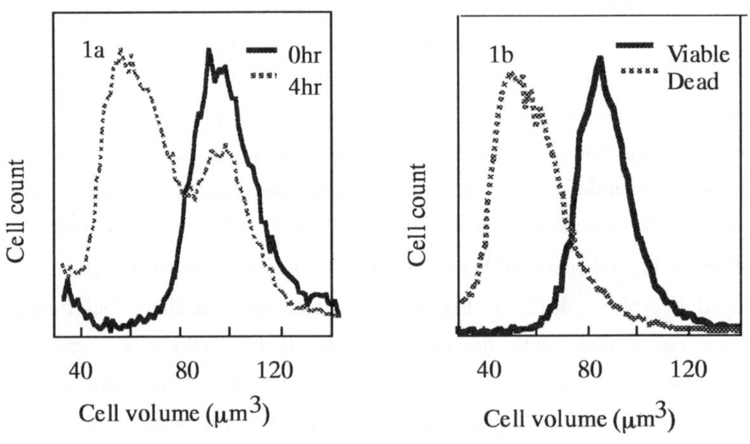

Fig. 1 Cell size reduction associated with radiation-induced apoptosis. 1a represents cell size distribution pattern of rat thymocytes measured at 0 and 4 hr after 10 Gy X-irradiation. Apoptotic (dead) thymocytes (1b) separated on Percoll gradient 3 hr after 10 Gy irradiation showed distinct reduction of cell size corresponding exactly to the small sized cell peak appeared after the irradiation shown in Fig. 1a.

1.3 Changes in Cell Surface Morphology

Scanning electron microscopic observations have shown that viable thymocytes possess microvilli on their surface. In contrast, the apoptotic thymocytes, having a higher buoyant density and reduced size, lost microvilli completely, and the surface was either smooth or extensively blistered[2].

Thus the apoptotic thymocytes are morphologically distinct from normal viable cells. No cells with intermediate morphological changes were detected, suggesting that transition from the viable to apoptotic state proceeds very quickly[5, 6].

1.4 Chromatin DNA Fragmentation

A rapid release of low molecular weight DNA from irradiated thymocytes has been known to be one of the major biochemical events associated with interphase death. We obtained in 1981 direct experimental evidence that the released DNA(-protein complex) consisted of integer multiples of a subunit, i.e., nucleosomal DNA[4]. Gel electrophoresis of the DNA showed a so-called "ladder pattern. "

Chromatin DNA fragmentation in individual cells could be demonstrated by single cell (micro) gel electrophoresis (tear-drop assay). Cells were embedded in gels on slide, and electrophoresed after lysis. Fragmented DNA in individual cells could be visualized as a tear-drop shaped migration pattern by fluorescent microscopic observation (Fig. 2). Here again, no intermediate pattern could be observed, suggesting that the apoptotic execution including DNA cleavage proceeds quickly in a short time.

Further support for this view was obtained from an electron microscopic observation of the thymocyte chromatin fibers[4]. Chromatin fibers of a normal thymocyte nucleus formed a network of chromatin strands. In contrast, chromatin fiber net work diminished from the samples prepared from apoptotic cells. Instead, a number of densely stained particles most possibly corresponding to oligonucleosomal fragments were observed.

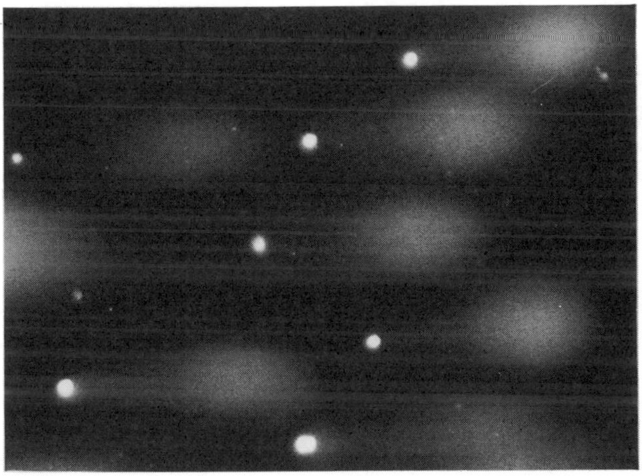

Fig. 2 Microgel electrophoretic detection of DNA fragmentation patterns in individual apoptotic cells. Apoptotic thymocyte population separated on Percoll gradient were embedded in gels on slide and subjected to electrophoresis after lysis. Fragmented chromatin DNA in the individual cells visualized by fluorescent microscopy showed migration pattern with tear-drop shape.

Endonuclease involved in the radiation apoptosis was purified from thymocytes and characterized[10], this will be reported in details in near future.

2. Peroxide Formation Associated with Radiation-induced Apoptosis

There is considerable evidence to show that an increase in reactive oxygen species (ROS) constitutes an intracellular signal leading to apoptosis[11]. Several antioxidants exert protective effects against the radiation-induced apoptosis in thymocytes.

In this study, we examined changes in peroxides, a kind of ROS, in X-irradiated rat thymocytes to define possible correlation between radiation-induced apoptosis and ROS. Intracellular peroxides were measured using 5-(and 6)-carboxy-2',7'-dichlorofluorescein diacetate (CDCFH-DA) and quantified with flow cytometry[12].

2.1. Peroxide Production during Apoptosis

Peroxide content in individual thymocytes was low immediately after X-irradiation. Cells with increased peroxide content began to appear at 2 hr. Subsequently, cells having remarkably higher peroxide content increased at 4 hr and simultaneously low peroxide cells decreased as shown in Fig. 3.

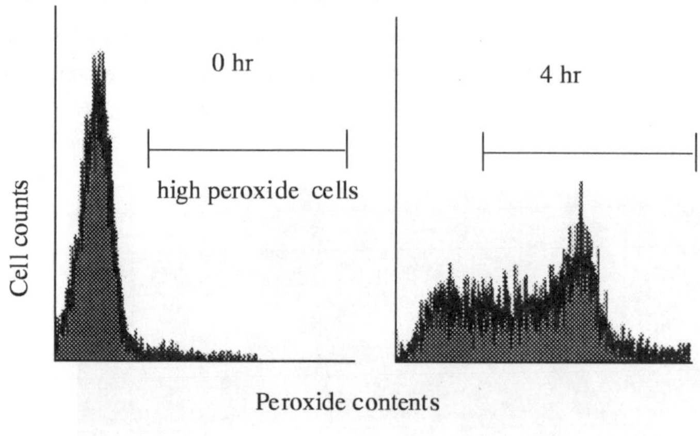

Fig. 3 Increase in number of cells with higher peroxide contents following 10 Gy X-irradiation. After 10 Gy X-irradiation peroxide contents in individual thymocytes were determined by flow cytometry.

As shown in Fig. 3, we defined and scored the cells with "high peroxide cells." The number of cells with high peroxide contents increased gradually with time after the irradiation in parallel with the increase in apoptotic cells showing chromatin condensation.

Flow cytometic examination revealed that the cells having normal cell volume showed lower peroxide levels, while the apoptotic cells with smaller sizes had higher peroxide contents.

The relationship between apoptosis and peroxide formation was further confirmed by determination of peroxide levels after separating apoptotic from preapoptotic viable cells by Percoll density gradient centrifugation technique. The results demonstrate clearly that the majority of viable cells contained lower peroxide, while most apoptotic cells had higher peroxide content.

All these findings indicate that the peroxides are not the primary products of irradiation, but are produced in parallel with the ongoing process of apoptotic executional phase in irradiated thymocytes.

Similar peroxide formation was observed during glucocorticoid-induced apoptosis in thymocytes (data not shown), providing further support for the view mentioned above.

2.2. Inhibition of Radiation-induced Apoptosis by Antioxidants

Several antioxidants and inhibitors exert protective effects against radiation-induced cell size reduction and peroxide formation. One of the newly synthesized antioxidant THCL (hydroquinone derivative, 2',4'-3,4-tetrahydroxychalcone) was found to have remarkable protective effects (Fig. 4).

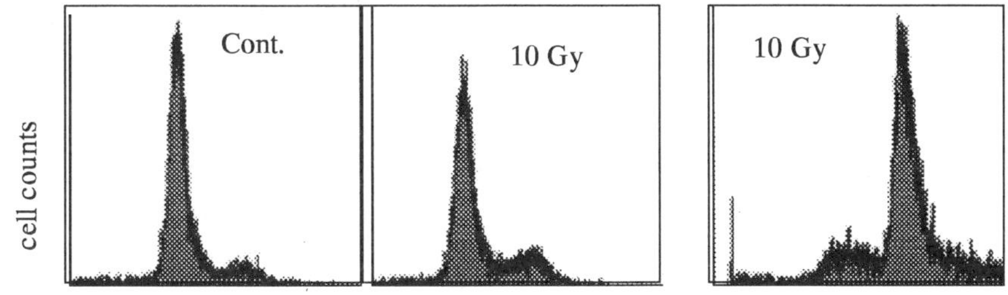

Fig.4 Prevention of peroxide formation and cell volume reduction by antioxidant THCL. Thymocytes exposed to 10 Gy X-irradiation were incubated for 4 hr in the presence of 0.1 mM THCL. Subsequently, cell volume (right column) and peroxide contents (left 2 column) in individual cells were determined by flow cytometry.

Both radiation-induced apoptotic peroxide formation and cell size reduction were almost completely prevented by addition of THCL. Similar protective effects were obtained by the addition of TEMPOL (free radical spin trapping agent), actinomycin D (inhibitor of RNA synthesis), and cycloheximide (inhibitor of protein synthesis). Several antioxidants such as glutathione and vitamin E, which did not prevent apoptosis, could not inhibit peroxide formation.

These findings strongly suggests that peroxide formation is closely related to the execution process in apoptosis.

2.3. Decrease in Glutathione(GSH+GSSG) during Apoptosis.

Protection of thymocytes from apoptosis by antioxidant suggests that some form of intracellular oxidation may play a role in radiation-induced apoptosis. So we attempted to examine changes in level of glutathione (GSH) which is one of the most abundant reductants present in mammalian cells.

Flow cytometric determination of intracellular glutathione revealed a close correlation between glutathione level and apoptosis (Fig. 5). Namely, most of the viable cells with normal cell size possessed higher glutathione level, while apoptotic cells of smaller size had lower glutathione content. To confirm the relationship between glutathione depletion and apoptosis, we again examined glutathione contents in apoptotic cells separated by Percoll gradient centrifugation. Biochemical determination of the total glutathione content revealed that a marked depletion of glutathione was found exclusively in apoptotic cells (data not shown). The present results suggest that the depletion of glutathione occurs parallel with the formation of peroxides in the process of radiation-induced apoptosis in thymocytes.

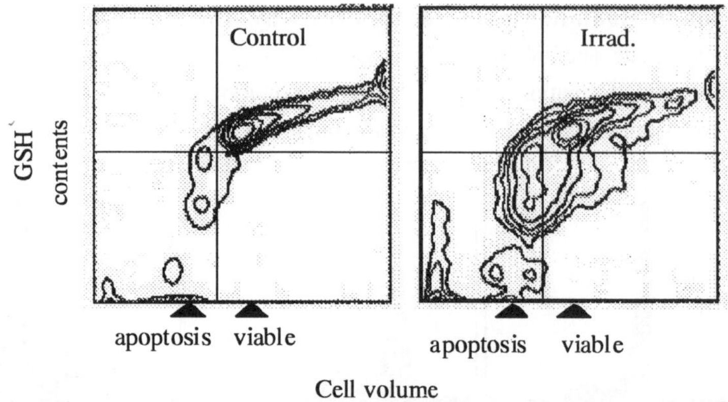

Fig.5 Depletion of glutathione in association with apoptosis. Glutathione content in individual control and 10 Gy X-irradiated thymocytes at 4 hr incubation was flow cytometrically determined after loading 5-chloromethylfluorescein diacetate(CMF-DA)[13].

Slater *et al.* reported previously that, when rat thymocytes were exposed to the glucocorticoid, a progressive increase in intracellular peroxides and a decrease in glutathione were spectrophotometrically observed to accompany the onset of apoptosis[14]. In the present study, we also observed an increase in peroxides and a reduction in glutathione in irradiated thymocytes using flow cytometric method. The method enabled us to examine directly the correlation between changes in peroxides or glutathione and the progression from viable to apoptotic state, which occurred simultaneously in individual cells.

These results suggest that the increase in peroxide constitutes a parallel step of the onset of common apoptotic execution phase, but not primary products of the irradiation that precedes and causes apoptosis.

The mechanisms for the peroxide formation and glutathione depletion remain to be elucidated. Reduced glutathione has been hypothesized to play a role in the rescue of cells from apoptosis[14]. Recently, none oxidative loss of glutathione in apoptosis via glutathione extrusion has been reported. The reduction in glutathione may have more general significance in the apoptosis. The peroxide formation observed in the present study may play some role in biodefence mechanisms against various stresses including radiation exposure.

References

1. T. Yamada and H. Ohyama, *Int. J. Radiat. Biol.* **37**: 342-351 (1980).
2. T. Yamada and H. Ohyama,*Scanning Electron Microsc.* **IV**: 55-60 (1981).
3. H. Ohyama., T. Yamada and I Watanabe, *Radiat. Res.* **85**: 333-339 (1981)

4. T. Yamada and H. Ohyama, *Radiat. Res*. **85:** 544-553 (1981).
5. T. Yamada and H. Ohyama, *Int. J. Radiat. Biol.*. **53**: 65-75 (1988).
6. T. Yamada and H. Ohyama, in : *Proc. 10th Int. Congress Radiat. Res.*, Wuerzburg, Germany, 1995 (Eds. U. Hagen, D. Harder, H. Jung & C.Streffer). pp. 645-648. (1996).
7. J. F. R. Kerr, A. H. Wyllie, and A. R. Currie, *Br. J.Cancer* **26**, 239-257 (1972).
8. H. Ohyama. and T. Yamada,*Scanning Electron Microsc.* **V:**1339-1343, (1983).
9. H. Ohyama., T. Yamada., A. Ohkawa and I. Watanabe, *Radiat. Res.* **101**: 123-130, (1985).
10. D. Shiokawa, H. Ohyama, T. Yamada, K. Takahashi and S. Tanuma, *Eur.J. Biochem.* **226**: 23-30 (1994).
11. T. M. Buttke and P. A. Sandstrom, *Immunol. Today,* **15**:7-10 (1994) .
12. J. P. Robinson, W. O. Carter and P. K. Narayanan, *Methods in Cell Biol.* **41**: 437-447, 1994.
13. D. W. Hedley and S. Chow, *Cytometry,* **15**: 349-358 (1994).
14. A. R. G. Slater, D. Stefan, I. Nobel, E. Maellaro, J. Butanante, M. Kimland and T. Orrenius, *Biochem. J.* **386:** 771-778 (1995).

9. DNA Fragmentation during Apoptosis Is Catalyzed by DNase γ

Sei-ichi TANUMA

Department of Biochemistry, Faculty of Pharmaceutical Sciences,
Science University of Tokyo, Shinjuku-ku, Tokyo, Japan

Apoptosis is a mechanism for eliminating unwanted cells from the cell community of multicellular organisms [1-4]. Abnormalities in the regulation of apoptosis may play a part in the etiology of cancer, autoimmune disease, viral infection, AIDS, and neurodegenerative disorders [5-7]. Although apoptosis is a gene-regulated cell death initiated by a wide variety of signals, the molecular mechanisms by which it is executed are still unclear. The elucidation of the regulatory mechanisms of apoptosis is important for developing chemotherapeutic agents against apoptotic diseases.

Thymocytes are widely used in the study of the molecular mechanisms of apoptosis [2-8]. An intriguing molecular event associated with thymic apoptosis is chromosomal DNA fragmentation into nucleosomal units [8-12]. Morphological changes, such as shrinkage and apoptotic body formation, are accompanied by this DNA fragmentation [8-12]. These phenomena characteristic of apoptosis have been suggested to be triggered by a sustained increase in intracellular Ca^{2+}. [4, 13, 14] Thus a likely candidate for an endonuclease responsible for DNA fragmentation in apoptosis is Ca^{2+}-dependent endonuclease(s). [8-12] A variety of studies lend support to the involvement of this type of endonuclease in apoptosis; Ca^{2+} ionophores induce DNA fragmentation and cell death in thymocytes and Ca^{2+} chelators can prevent both processes. [11, 14] Furthermore, apoptosis in thymocytes is known to be inhibited by Zn^{2+}. [13] An apoptotic Ca^{2+}-dependent endonuclease is, therefore, thought to be sensitive to Zn^{2+}.

Previously, we found that three types of deoxyribonuclease (DNases α, β, and γ) were present in rat thymocytes nuclei. [15] These endonucleases, which are clearly separated on CM5PW HPLC, cleave chromatin to the nucleosomal fragments *in vitro* . [15] However, the functions of the three forms of nuclear DNase remain to be undetermined. Considering the central role of the endonuclease in thymic apoptosis, we have attempted to purify these three nuclear DNases and compared their properties with the nature of DNA fragmentation in apoptosis.

Results

We first examined the changes in activities of these three DNases simultaneously with changes in DNA fragmentation and cell morphology during apoptosis. After 3 hr irradiation, thymocytes exhibited typical morphological changes (Fig. 1a) and internucleosomal DNA fragmentation (Fig. 1b). The transition of a collapsed nucleus into several nuclear fragments was a rate limiting step, since the collapsed black hole nuclei were present in high frequency even after 6 hr (Fig. 1a). DNase activities present in normal and apoptotic rat thymocyte nuclei were almost completely solubilized by sonication at high ionic strength separated by CM5PW HPLC. Interestingly, the induction of apoptosis by irradiation resulted in decreases in the activities of DNases α and β, without affecting that of DNase γ appreciably (Fig. 1c). These changes in DNase activities were also observed during apoptosis induced by a synthetic glucocorticoid, dexamethasone.

Thus a putative apoptotic endonuclease, DNase γ from apoptotic nuclei and DNases α and β from normal nuclei were purified by sequential HPLC steps. On G2000SW gel filtration HPLC, DNases α, β, and γ appeared as single peaks of approximate 30 kDa. These purified DNases on a final heparin 5PW HPLC were homogeneous as judged by SDS-PAGE - silver staining (Fig. 2, inset lane 1). The *Mrs* of DNases α, β, and γ were calculated to be 32, 32 and 33 kDa, respectively. We confirmed that these purified endonuclease activities were restricted to the regions of stained protein bands by the SDS/DNA-PAGE - renaturation (activity gel) method. This activity gel method is based on the ability of DNase to be renaturated after removal of SDS and to cleave DNA during incubation of the gels under optimum conditions of the enzymes. As shown in Fig. 2, inset lane 2, the localization of DNase activities in gels can be detected by the disappearance of DNA fluorescence as dark bands on an ethidium bromide-fluorescent background. DNases α, β, and γ exhibited non-fluorescent bands corresponding to proteins of 32, 32, and 33 kDa, respectively.

DNase γ required both Ca^{2+} and Mg^{2+} for full activity (Fig. 3i and j). Of the other divalent cations tested, only Mn^{2+} partially substituted for Ca^{2+}/Mg^{2+} (Fig. 3k). The optimal concentrations of these divalent cations were 1 - 3 mM. The DNase γ was sensitive to Zn^{2+}, half-maximal inhibition being observed with a Zn^{2+} concentration as low as 40 μM (Fig. 3l). This Zn^{2+} sensitivity of DNase γ was consistent with the observations at the nuclear and cellular levels. In contrast, DNases α (Fig. 3a - d) and β (Fig. 3e - h) were not affected by these divalent cations. These results suggest that DNase γ present in apoptotic rat thymocyte nuclei catalyzes DNA fragmentation during thymic apoptosis.

To confirm this possibility, we compared the mode of DNA hydrolysis by DNase γ with that in apoptotic cells (Fig. 4). DNA endo-labeling methods were used for

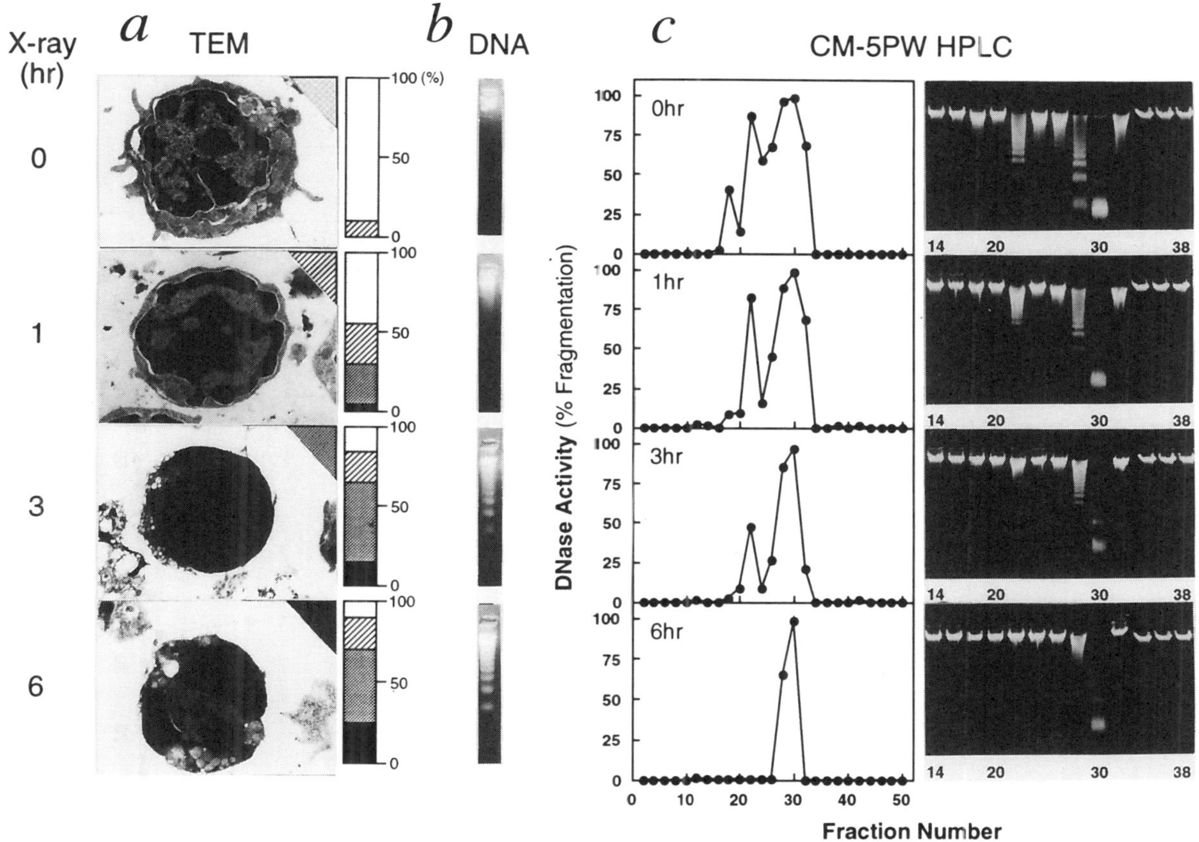

Fig. 1. Morphological and biochemical changes in apoptosis of X-irradiated rat thymocytes. Morphological changes of rat thymocytes after exposure to X-rays (10 Gy) were observed by transmission electron microscopy (TEM) (x35,000 magnification) Typical electron micrographs observed at intervals are illustrated and their frequencies are shown by histograms (a). DNA fragments extracted from normal and apoptotic thymocytes were analyzed by 2% agarose gel electrophoresis (DNA) (b). Changes in activities of DNases α, β and γ in nuclei were analysed by CM5PW HPLC (right, HeLa S3 nuclear assay - agarose gel electrophoresis; left, estimation of % fragmentation) (c).

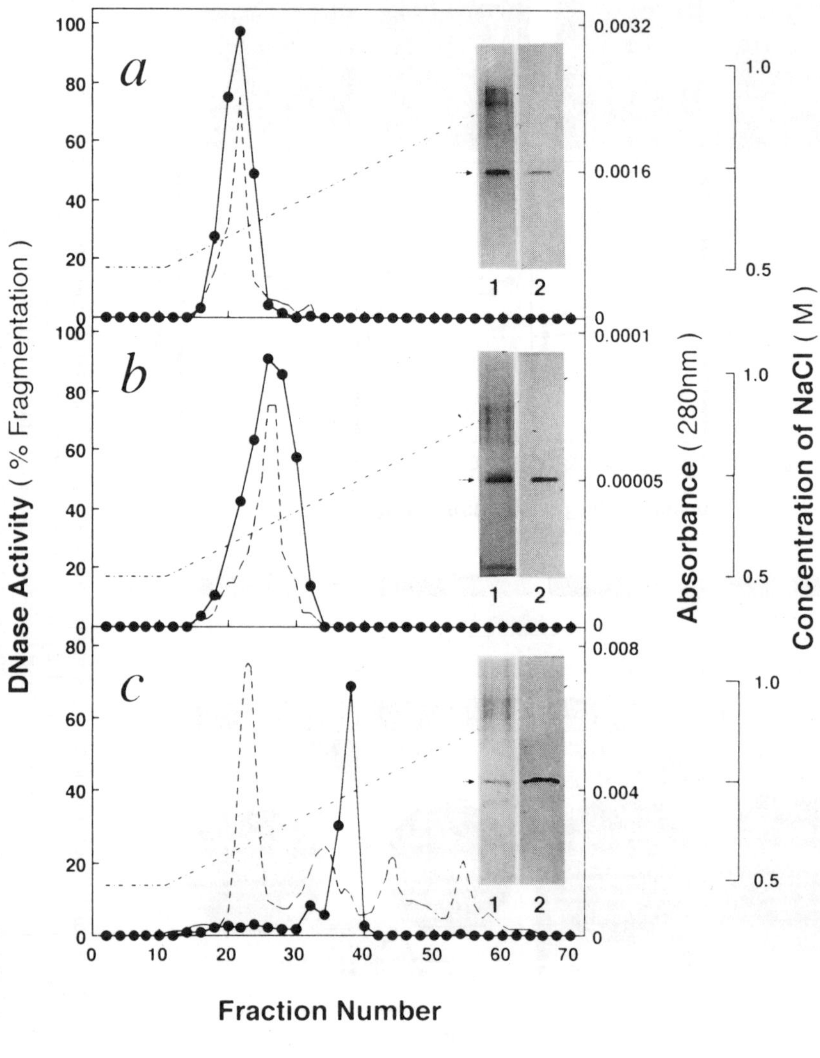

Fig. 2. Purification of DNases α, β, and γ on heparin5PW HPLC. Elution profiles on a final heparin5PW HPLC of DNases α (a), b, (b) and γ (c). The insets indicate SDS-PAGE - silver staining (lane 1) and activity gel analysis (lane 2) of DNases α, β, and γ from the peak fractions.

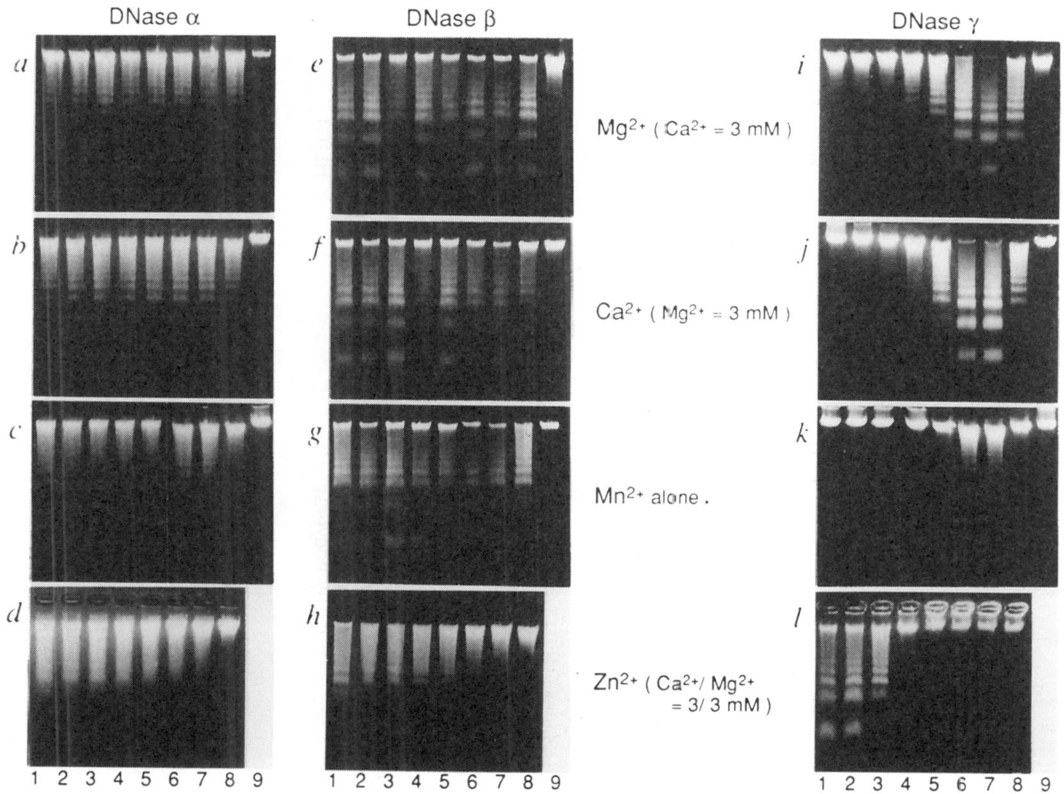

Fig. 3. The effects of divalent cations on purified DNases α, β, and g. The activities of purified DNases a (a - d), β (e - h), and g (i - l) were assayed in the presence of various concentrations of Mg^{2+} (a, e, i) Ca^{2+} (b, f, j), Mn^{2+} (c, g, k), or Zn^{2+} (d, h, l) under standard assay conditions for each DNase. The purified DNase activities were assayed in the presence of 3 mM $CaCl_2$ and increasing concentrations of $MgCl_2$ (a, e, i), in the presence of 3 mM $MgCl_2$ and increasing concentrations of $CaCl_2$ (b, f, j), in the presence of $MnCl_2$ alone (c, g, k), or in the presence of various concentrations of $ZnCl_2$ (d, h, l) under optimum conditions for each DNase. The concentrations of $MgCl_2$, $CaCl_2$ and $MnCl_2$ were 0, 0.01, 0.03, 0.1, 0.3, 1, 3, 10 and 30 mM (lanes 1 - 9), and those of $ZnCl_2$ were 0, 0.01, 0.03, 0.1, 0.3, 1, 3 and 10 mM (lanes 1 - 8).

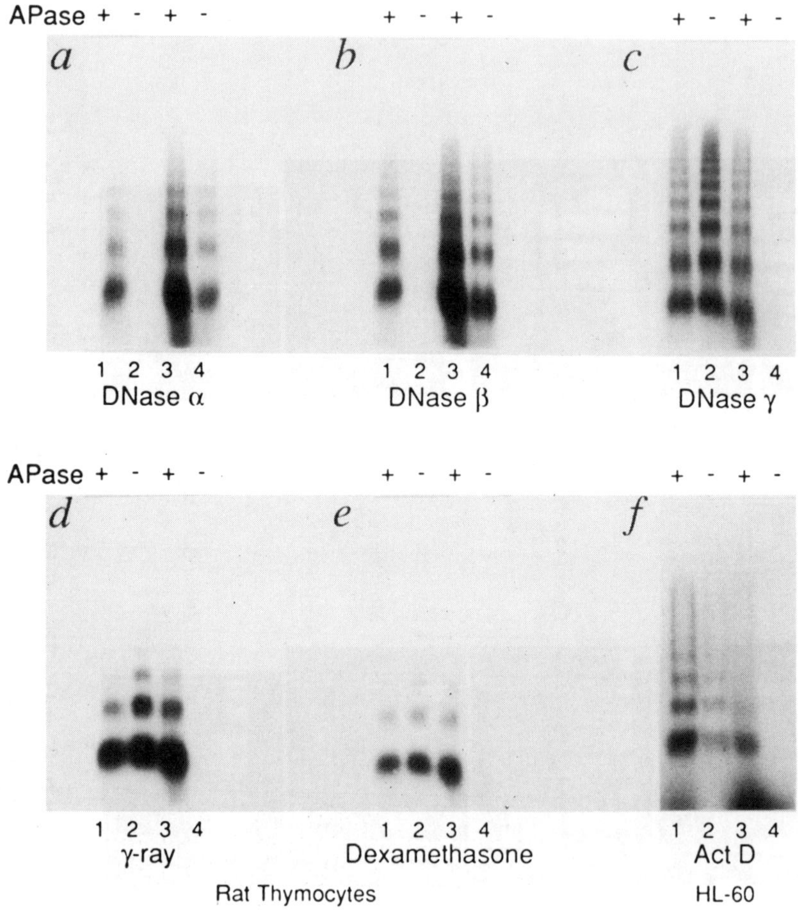

Fig. 4. Modes of DNA cleavage by DNases α, β, and γ, and in apoptotic rat thymocytes. DNA extracted from HeLa S3 cell nuclei digested by DNases α (a), β (b), or γ (c), and from apoptotic rat thymocytes induced by irradiation (10 Gy) (d) or dexamethasone (10^{-7} M) treatment (e) for 4 hr and apoptotic HL-60 cells induced by Act D (1 μg/ml) for 4 hr (f), was subjected to 3' end- (lanes 1 and 3) or 5' end- (lanes 3 and 4) labelling after pretreatment with (lanes 1 and 2) or without (lanes 2 and 4) APase. Aliquots of DNA were analyzed by 2% agarose gel electrophoresis and autoradiography.

determination of the modes of DNase actions. If DNA fragments have free 3'-OH and 5'-*P* ends, the resulting nucleosome ladders should be detected simply [without alkaline phosphatase (APase) pretreatment] by 3' end-labeling of the extracted DNA with terminal deoxynucleotidyl transferase (TdT) and [α-^{32}P]dCTP, and by 5' end-labeling with polynucleotide kinase (PNK) and [γ-^{32}P]ATP only after pretreatment of the DNA with APase. In the opposite case (3'-*P*/5'-OH ends), only 3'-ends of DNA fragments should not be labeled without APase pretreatment. Purified DNA from HeLa S3 cell nuclei digested with DNase γ was end-labeled at 3' ends with TdT and [a-^{32}P]dCTP without APase pretreatment, but the 5' ends of the fragments were not labeled with PNK and [γ-^{32}P]ATP without APase pretreatment. Thus, DNase γ produced DNA fragments with 3'-OH/5'-*P* ends. In contrast, DNase α or β produced fragments with 3'-*P*/5'-OH ends, as evidenced by labeling of the 3' ends only after APase pretreatment and 5' end-labeling without APase pretreatment.

The nature of DNA fragmentation at the cellular level was examined by three different systems (Fig. 4): i) irradiation of rat thymocytes with X-rays, ii) incubation of rat thymocytes in the presence of dexamethasone, and iii) incubation of human promyelocytic leukemia (HL-60) cells in the presence of actinomycin D (Act D). These treatments triggered extensive DNA fragmentation prior to cell death. The resultant autoradiograms of endo-labeled DNA fragments revealed that the 5' ends of the DNA fragments could not be labeled without APase pretreatment, suggesting that in all cases, the fragments formed had 3'-OH/5'-*P* ends. These labeling patterns were the same as those obtained with DNase γ.

Discussion

In this study we identified and purified a Ca^{2+}/Mg^{2+}-dependent and Zn^{2+}-inhibitable endonuclease, DNase γ, from rat thymocytes. Judging from its physical and catalytic properties, DNase γ differed from NUC18 (*Mr* = 18 kDa), [16] DNase I (*Mr* = 31 kDa) [17] and CHO cell acidic endonuclease (*Mr* = 30 and 31 kDa, doublet) (DNase II), [18] which were recently reported to be putative apoptotic endonucleases. Its properties also distinguished it from recently purified mammalian Ca^{2+}/Mg^{2+} -dependent (*Mr* = 27 kDa) [19] and Ca^{2+}/Mn^{2+} -dependent (*Mr* = 22 kDa) [20] endonucleases. More significantly, the present study showed that the cleavage ends of DNA fragments in apoptotic cells and those produced by DNase γ were the same. The similar DNase γ activity was purified from calf thymocytes and HL-60 cells (unpublished data). Thus, we conclude that DNase γ is an important component of the apoptotic machinery in thymocytes. The activities of the cognate form of DNases α and β were changed by various stimuli (unpublished results). Although the physiological functions of these acidic endonucleases are still unknown, they may be important in some nuclear functions for cell survival. Elucidation of the mechanisms activating these three nuclear DNases and their gene regulation may provide important clues for understanding their biological significance in apoptotic cell death or cell survival.

There are several possible explanations for the physiological significance of DNA fragmentation. First, selective DNA degradation may occur at specific DNA domains involved in cell proliferation, such as replication origins and telomeres. Second, integrated c-oncogenes and proviral genes may be cleaved for prevention of transformation by their transfection. Third, protection from liberation of large amounts of DNA causing autoimmune reactions is necessary, since antibodies against double-stranded DNA are associated with SLE. Fourth, the formation of apoptotic bodies leading to DNA fragmentation facilitates their engulfment by phagocytes. Although speculative, these explanations indicate the possible importance of DNA fragmentation during apoptosis and suggest ways in which it may have very distinct spatial and temporal significance.[21] Ultimately, the purpose of apoptosis is considered to be elimination of individual cells aged and damaged sufficiently to increase risk of heritable genetic changes, and maintenance of genetic fidelity. Thus it is principally a genetic program for gene survival *via* gene elimination.[22]

Acknowledgments

We thank Prof. T. Yamada (Toho University) and Dr. H. Ohyama (National Institute of Radiological sciences) for helpful suggestions. This work was supported in part by a Grant-in-Aid for Scientific Research from the Ministry of Education, Science and Culture of Japan.

References

1. Kerr, J. F. R., Wyllie, A. H., and Currie, A. R. *Br. J .Cancer*, **26**, 239-257 (1972)
2. Umansky, S. R., *J. Theor. Biol.*, **97**, 591-602 (1982)
3. Raff, M. C., *Nature* , **356**, 397-400 (1992)
4. Cohen, J. J., *Immunol. Today* , **14**, 126-130 (1993)
5. Levine, A. J., Momzd, J. and Finlay, C. A., *Nature*, **351**, 453 -456 (1991)
6. Carson, D. A. and Ribeiro, J. M., *Lancet,* **341**, 1251 - 1254 (1993)
7. Cougeon, M L., and Mantagnier, L., *Science*, **260**, 1269 - 1270 (1993)
8. Wyllie, A. H., *Nature*, **284**, 555-556 (1980)
9. Wyllie, A. H., Morris, R. G., Smith A. L., and Dunlop, D., *J. Phthol.*, **142**, 67-77 (1984)
10. Arends, M. J., Morris, R. G., and Wyllie, A. H., *Am. J. Pathol.*, **136**, 593-601 (1990)
11. Peitsh, M. C., Mannherz, H. G., and Tschopp, J., *Trends Cell Biol.*, 4, 37-41 (1994)

12. Tanuma, S,. and Shikawa, D., *Biochem. Biophys. Res. Commun.,* **203**, 789-797 (1994)
13. Cohen J. J. and Duke, R. C., *J. Immunol.*, **132**, 38 (1984)
14. McConkey, D. J. Hartzell, P. Nicotera P., and Orrenius, S., *FASEB J.*, **3**, 1843 (1989)
16. Gaido, M. L., and Cidlowski, J. A., *J. Biol. Chem.*, **266**, 18580-18585 (1991)
15. Shiokawa, D. Ohyama, H. Yamada, T. Takahashi, K., and Tanuma, S., *Eur. J. Biochem.,* **226**, 23 (1994).
17. Peitsch, M. C., Polzar, B., Stepham, H., Crompton, T., MacDonald, H. R., Mannherz, H. G., and Tshopp, J., *EMBO J.*, **12**, 371-377 (1993)
18. Barry, A., and Eastman, A., *Arch. Biochem. Biophys.*, **300**, 440-450 (1993)
19. Ribeiro, J. M., and Carson, D. A., *Biochemistry*, **32**, 9129-936 (1993)
20. Nikonova, L. V., Belelsky, I. P., and Umansky, S. R., *Eur. J. Biochem.*, **215**, 893-901 (1993)
21. Tanuma, S., *Tissu. Cult. Res. Commun.*, **14**, 133-140 (1995)
22. Tanuma, S. In: Apoptosis in Normal Development and Cancer (Ed. by Sluyser, M.) Taylor and Francis Ltd., London UK, pp. 39-59 (1996)

10. Biodefence Mechanisms against Radiation-induced Apoptosis

ZVI FUKS
Department of Radiation Oncology, Memorial Sloan-Kettering Cancer Center, New York, USA

I. Introduction

Exposure of mammalian cells to adverse environmental conditions leads to induction of stress responses, associated with activation of specific genes or with alterations of constitutively functioning genes. One of the fundamental responses is the activation of a programmed pathway of cell death, known as apoptosis, which represents a prevalent mechanism for removing cells injured by stress. Apoptosis is an inducible signal transduction pathway of biochemical events, the elements of which are constitutively expressed in every cell, albeit in an inactive form (1, 2). Induction of apoptosis is increasingly identified in response to stress *in vitro* and *in vivo*, but its expression is frequently suppressed by anti-apoptotic mechanisms that protect the cell against excessive and uncontrolled cell death in response to repairable or non-injurious stress stimuli (3). The notion that stress-induced apoptosis may by tightly regulated by anti-apoptotic mechanisms, and that the balance between pro- and anti-apoptotic systems may determine under any set of circumstances the magnitude and prevalence of the apoptotic response, is the topic of this review. Radiation-induced apoptosis and the cellular mechanisms that regulate this response will be described as a model for the interplay between environmental stresses and the biodefence mechanisms that control the evolution of the apoptotic response. Improved understanding of the coordinated function of pro- and anti-apoptotic signaling systems may yield opportunities to explore pharmacological manipulations to enhance or restrain the apoptotic response in cells and tissues exposed to potentially damaging environmental stresses.

II. Signal transduction pathways in radiation-induced apoptosis

Although apoptosis is increasingly identified as a radiation response in mammalian cell systems, it is still regarded as a rare primary mode of radiation-induced cell kill and as a minor component of the therapeutic action of radiation (3). The most common mode of radiation-induced cell kill is the mitosis-dependent (also known as the reproductive or clonogenic) mechanism (4, 5). The lethal action in this model involves the cell nucleus as the target, DNA double-strand breaks as the specific lesions, and genetic instability associated with unrepaired or misrepaired DNA breaks as the mechanism that leads to cell death (4, 5). Lethal mutations or dysfunctional chromosomal aberrations are believed to lead to progeny cell death, usually after several mitotic cycles (4-6). In contrast, apoptotic simuli such as cytokines (i.e. TNF, Fas), antibodies, or environmental stresses (i.e ionizing ionizing and UV radiation, heat shock, oxidative stress) appear to target cell surface receptors to trigger apoptosis (3, 7). Further, while the signaling pathway initiated may differ according to the type of stimulus and the targeted receptor, the various upstream cascades converge downstream to a common final effector mechanism for disassembling the dying cell into subcellular structures, which involves the ICE/Ced-3 family of cysteine proteases (8). Preservation of the integrity of the plasma membrane until late in the process enables packaging of disintegrating organelles into membrane-bound vesicles, which are phagocytized by neighboring cells or macrophages without inducing a reactive inflammatory tissue response. The net result is that lethally damaged cells are removed without producing tissue damage.

Studies carried out over the past three years have identified the cell membrane targets for environmental stresses in initiating apoptosis and characterized several of the upstream biochemical events which mediate the transfer of apoptotic signals to the cytosol and nucleus. These studies showed that ionizing and UV radiation, heat shock and oxidative stress utilize the sphingomyelin signal transduction pathway to initiate apoptosis (9, 10). The sphingomyelin pathway is a ubiquitous, evolutionary conserved signaling system, which is initiated by hydrolysis of the phospholipid sphingomyelin (N-acylsphingosine-1-phosphocholine) via the action of sphingomyelin-specific forms of phospholipase C, termed sphingomyelinases (11). Sphingomyelin is preferentially concentrated in the outer leaflet of the plasma membrane of mammalian cells, and radiation was shown to activate membrane sphingomyelinase to hydrolyze the phosphodiester bond of sphingomyelin to yield ceramide and phosphorylcholine (11, 12). Ceramide then serves as a second messenger in activating the cascade of biochemical events which transfer signals to the cytoplasm and the nucleus sequentially to execute apoptosis.

The role of ceramide as a second messenger in activating the apoptotic signal transduction pathway was initially suggested three years ago in studies on the mechanism of the apoptotic action of TNFα (13, 14). Haimovitz-Friedman et al. (9) reported shortly afterwards that ionizing radiation activated cell membrane sphingomyelinase to generate ceramide and initiate apoptosis in primary cultures of bovine aortic endothelial cells (BAEC). Exposure of BAEC to ionizing radiation induced within seconds a dose-dependent sphingomyelin hydrolysis to generate ceramide (ED_{50} of 1.5 Gy) and apoptosis (LD_{50} of 3 Gy). Figure 1 shows that exposure of BAEC to a dose of 500 cGy increased cellular ceramide from 0.95 to 1.3 pmol/10^6 cells ($p <$ 0.025) within 30 seconds of irradiation. Rdaiation mimicked the apoptotic effect of ceramide

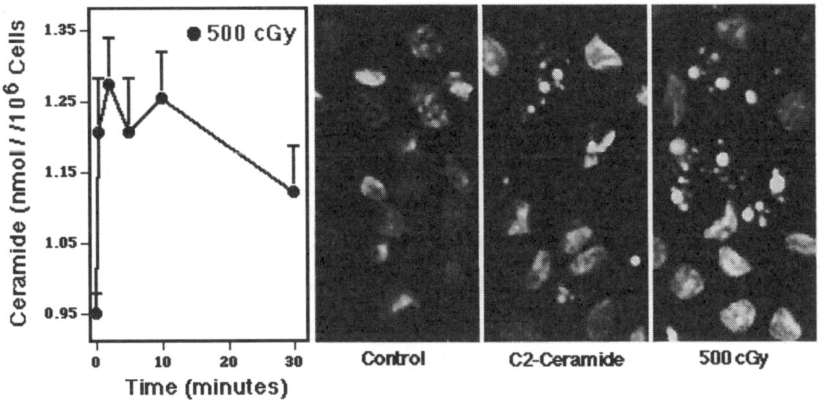

Figure 1. Radiation Induces Ceramide Generation and Apoptosis in BAEC. The left panel shows a time course of ceramide generation in BAEC exposed to a dose of 500 cGy. Ceramide was quantified by the DAG kinase assay as described (9). The right panels show typical morphological features of apoptosis in the nuclei of BAEC treated with the C2-ceramide analog (50µM, 24 hrs) or in cells exposed to 500 cGy. Apoptosis was assessed by staining with the DNA-specific fluorochrome bisbenzamide. Radiation is shown to mimic the apoptotic effect of C2-ceramide

analogs (C2- and C6-ceramide) (Figure 1). These studies also showed that PKC activation by TPA completely blocked radiation-induced sphingomyelinase activation and the resulting generation of ceramide and apoptosis. However, apoptosis could be restored in TPA treated cells by C2-ceramide added exogenously, strongly suggesting that ceramide is indeed an essential activator of the apoptotic pathway. Ionizing radiation (10 Gy) acted directly on BAEC membrane preparations devoid of nuclei, stimulating ceramide generation by activating the sphingomyelinase enzyme. The level of ceramide in irradiated membranes obtained from 10^6 cells increased within 4 minutes from a basal level of 2.25 to 2.75 nmol/mg supernate protein ($p<0.001$), a rise similar in magnitude to that observed after irradiation of a similar number of intact BAEC. These studies suggested that apoptotic signaling can be generated by interaction of ionizing radiation with cellular membranes and suggested an alternative to the model of DNA damage as the mediator of radiation-induced cell kill.

Definitive evidence for the role of ceramide in radiation-induced apoptosis was provided by experiments which used genetic models of acid sphingomyelinase deficiency (15). EBV-transformed normal human lymphoblasts responded to dose of 1-20 Gy with typical time- and dose-dependent decrease of cellular sphingomyelin, ceramide generation and apoptosis. In contrast, EBV-transformed lymphoblasts from Niemann-Pick disease (NPD) patients, who have an inherited deficiency of acid sphingomyelinase activity, failed to generate ceramide and to undergo apoptosis even when exposed to a dose of 20 Gy. These abnormalities were completely reversed upon restoration of acid sphingomyelinase activity by retroviral transfer of human acid sphingomyelinase cDNA. Experiments were also carried out with acid sphingomyelinase knock-out mice. Irradiated lungs from wild type mice showed a 2.3 fold increase in tissue ceramide level at 30 minutes after exposure to 20 Gy, and an extensive apoptotic response at 10 hours. In contrast, acid sphingomyelinase-knockout mice failed to show ceramide generation and apoptosis in the lung at any dose up to 30 Gy. Comparison of patterns of tissue apoptosis after whole body irradiation in acid sphingomyelinase- and p53-knockout mice suggested that acid sphingomyelinase-and p53-mediated apoptosis are most likely distinct and independent. While endothelial cells in the lung and the endocard, as well as mesothelial cells of the pleura and the pericard, utilized specifically the acid sphingomyelinase system to initiate apoptosis and were not affected by mutation of the p53 gene, T-cells of the thymus utilized mostly of the p53-dependent pathway and were not affected by mutation of the acid sphingomyelinase gene. These data suggested that radiation is capable of activating at least two independent signaling mechanisms for induction of apoptosis, and provided definitive evidence for the role of ceramide as a second messenger in initiating the sphingomyelinase-mediated pathway.

The transfer of apoptotic signals from the irradiated cell membrane to the cytosol appears to employ the stress-activated protein kinase (SAPK/JNK) cascade (10), involving sequential activation of MEKK1, SEK1, SAPK and c-Jun. Verheij et al. (10) showed that exposure of U937 cells or BAEC to ionizing radiation, hydrogen peroxide, ultraviolet C radiation, heat shock or TNFα induced within seconds to minutes significant elevations of ceramide and ceramide-dependent activation of the SAPK/JNK cascade, assessed by the ability of immunoprecipitated SAPK to phosphorylate the *N*-terminal c-Jun construct GST-c-Jun-1-135. Overexpression of a dominant negative mutant of c-Jun blocked stress- and ceramide-induced apoptosis, but did not compromise the ability of stress to elevate cellular ceramide or to activate SAPK. Similarly, overexpression of dominant negative SEK1 abolished stress- and ceramide-induced apoptosis. These studies provided definitive evidence for a coordinated regulation of radiation-induced apoptosis in BAEC and U937 cells via the ceramide and SAPK pathways. The involvement of SAPK/JNK in radiation-induced apoptosis was confirmed by Chen et al (16), in the human Jurkat, HL-60, BV173 cells and in 239T renal embryonic cells. In contrast, Nishina et al. (17) reported that thymocytes from SEK1 knockout mice showed a normal apoptotic sensitivity to UV and ionizing radiation, indicating that radiation may signal apoptosis via another as yet unknown signaling pathway. The mechanism by which the SAPK/JNK system links to ICE/Ced-3 proteases to trigger the final effector function of the apoptotic pathway remains unknown.

ICE/Ced-3 proteases appears, nonetheless, to be involved in the apoptotic response to radiation. Emoto et al. (18) reported that radiation induced in U937 cells PKCδ cleavage (a substrate for ICE-like proteases) and apoptosis, and that these effects were inhibited by the ICE-like protease inhibitor Y-VAD.

III. Protein Kinase C (PKC) signals anti-apoptotic control mechanisms

A variety of anti-apoptotic mechanisms have been described in the recent literature, including the Bcl-2 system, (19, 20) loss of p53 activity (21, 22), activation of NF-κB (23-25), growth factor stimulation (26, 27), and activation of DAG-PKC signaling system (9, 27-29). Substantial evidence has been provided by a variety of studies to indicate that activation of PKC serves as one of the main anti-apoptotic mechanisms against ceramide-mediated apoptosis Phorbol esters, shown to be anti-apoptotic in diverse systems, was found to antagonize ceramide-mediated apoptosis induced by TNF (13, 29). Similarly, basic fibroblast growth factor (bFGF) was reported to inhibit radiation-induced apoptosis in BAEC via activation of the DAG-PKC pathway (26, 27). PKC activation by the phorbol ester TPA (30 ng/ml) mimicked the anti-apoptotic effect of bFGF. Agarose gel electrophoresis showed that both TPA and bFGF inhibited the apoptotic degradation of DNA induced by exposure to 5 Gy. The anti-apoptotic effect of bFGF was abrogated by pre-treatment with the PKC inhibitor H 7 or by depletion of cellular PKC by overnight preincubation with high dose TPA.

Studies on the mechanism of the anti-apoptotic effects of PKC disclosed that phorbol ester-activated PKC inhibited radiation-induced sphingomyelinase activation and ceramide generation (9), thus defining an upstream anti-apoptotic checkpoint at the sphingomyelinase level. In some cells, this effect may be sufficient to abrogate apoptosis. However, PKC appears to control additional anti-apoptotic checkpoints along the ceramide-mediated apoptotic pathway. Consistent with this notion are the reports by Jarvis et al.(29), who showed that PKC activation abrogated apoptosis in response to treatment with ceramide analogs, indicating that PKC is capable of inhibiting apoptosis at checkpoints downstream of ceramide generation. One such critical anti-apoptotic checkpoint controls the pro-apoptotic function of the SAPK/JNK cascade. Xia et al. (30) reported that withdrawal of nerve growth factor (NGF) in PC-12 cells resulted in activation of the SAPK/JNK and p38 signaling systems, and in concomitant reduction in the activity of the extracellular regulated kinase (ERK) cascade. In contrast, constitutive activation of ERK down-regulated SAPK/JNK and prevented apoptosis.

Although in some mammalian cells, growth factor or TPA stimulation is required to elicit the activation of PKC, other cells are capable of responding to radiation with direct activation of PKC, which may provide a mechanism for radiation resistance in some cells (31-34). Hallahan *et al.* reported activation of PKC within 1 minute after radiation exposure of HL-60 (31) and RIT-3 cells (32), while Kim *et al* (33) found that radiation activated the ε isoform of PKC in human lung adenocarcinoma A549 cells. These studies indicate that anti- and pro-apoptotic signaling can be co-regulated by the same stress. A similar phenomenon was reported for NF-κB activation. TNFα, ionizing radiation, or daunorubicin were reported to activate NF-κB which signals anti-apoptosis, attenuating the pro-apoptotic response to these stresses (23-25).

The studies on the mechanisms of anti-apoptotic signaling that regulate ceramide-mediated apoptosis collectively indicate that while ceramide initiates the apoptotic pathway, it does not commit the cell to the death process. PKC signaling can reverse the process at multiple check points along the apoptotic signaling pathway. The studies also confirm that the balance between pro- and anti-apoptotic signaling, regulated through the two lipid second messengers ceramide and DAG, can be modulated pharmacologically. Shifting the balance to enhance either the pro- or the anti-apoptotic phenotype may affect the level of apoptotic cell kill by radiation, representing an approach to modulate the effects of radiation in normal and tumor cells with a potential for clinical applications.

IV. The Anti-apoptotic Effects of bFGF *in vivo*

To test whether the model of reciprocal regulation of apoptosis via ceramide and PKC is also relevant to *in vivo* systems, studies were carried out with a mouse lung model (26, 35). The lung was selected as a target because it is enriched with a network of capillaries, and radiation exposure of the lung results in a lethal syndrome, known as radiation pneumonitis. It has been suggested that radiation-induced damage to pulmonary microvascular endothelial cells serves as a basic and essential element in the pathogenesis of lethal radiation pneumonitis (35). Experiments with several strains of mice exhibited radiation-induced time- and dose-dependent ceramide generation in the lung tissue (15). C3H/HeJ and C3H/scid mice receiving whole chest irradiation exhibited extensive apoptotic changes in the endothelium of the capillary network of the lung, in the endocard, and in the mesothelial cells of the pleura and the pericard (35). Quantitative evaluation of TUNEL-positive cells in the irradiated lung revealed a time- and dose-dependent increase in endothelial cell apoptosis in the lung, reaching a level 29.7% at 10 hours after exposure to 20 Gy, compater to a background level of 2.4% in untreated animals ($p>0.0001$) (35). The apoptotic response to radiation was nearly completely inhibited by intravenous injection of recombinant bFGF, with only occasional endothelial cells showing evidence of apoptosis (26, 35). A special characteristic of intravenously injected bFGF is that bFGF does not extravasate through the blood vessel wall and remains bound to the intravascular surface of the endothelium, thus determining the patterns of the anti-apoptotic effects of bFGF therapy (26). Thus, in addition to inhibiting radiation-induced apoptosis in microvascular endothelial cells, intravenous bFGF also inhibited the apoptotic response in the endocardium, but had no effect on the apoptotic changes in thymocytes, and in the pleural and pericardial mesothelial cells (35).

The demonstration that bFGF remains trapped in the vascular intima *in vivo* and abrogates radiation-induced apoptosis suggested that bFGF treatment might provide a protection of the pulmonary parenchyma against radiation-induced damage (35). Figure 2 shows that treatment with intravenous bFGF inded rescued mice from lethal radiation pneumonitis. When radiation

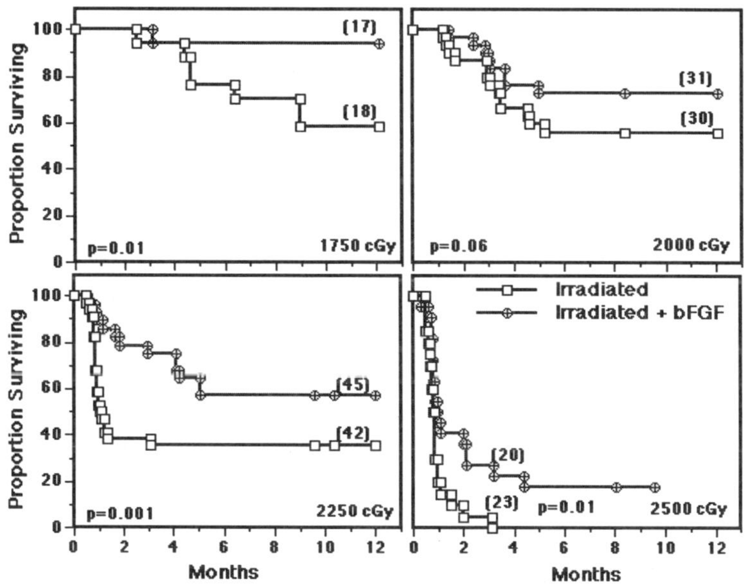

Figure 2. Basic FGF Protects C3H/HeJ Mice Against Lethal Radiation Pneumonotis. Mice were exposed to bilateral lung irradiation using the technique of whole chest irradiation with cerrobend blocks shielding the mediastinum. Actuarial survival curves are shown for mice exposed to single doses of radiation as indicated, with or without recombinant bFGF (100 ng x 4) injected intravenously immediately before and after irradiation and at 1 and 2 hours later. The number in parenthesis indicate the number of animals treated in each group

was restricted to the lungs (delivered via a whole chest radiation technique with a mediastinal cerrobend block to shield the esophagus and heart), the $LD_{50/180}$ from radiation pneumonitis in C3H/HeJ mice was 20.75 Gy and it increased to 23.0 Gy in bFGF treated mice (35). The radioprotective effect of bFGF was significantly more pronounced at the low dose range, below the LD $_{50/180}$, than at higher doses.

These studies provided a proof in principal that apoptosis may represent a biologically relevant mechanism of radiation-induced cell kill *in vivo*, and that protection against this effect may be associated with the level of radiation sensitivity in irradiated tissues.

V. Summary

The mechanisms of pro- and anti-apoptotic signaling via ceramide and PKC reviewed here provide a model for the study of mechanisms of cell injury by environmental stresses and the biodefence mechanisms that protect the organism against aberrant or overexpression of lethal responses. The sphingomyelin system serves a ubiquitous mechanism for transmembrane signaling leading to cell death, but bioprotective mechanisms mediated via PKC tightly regulate this apoptotic response, and the "cross-talk" between these pro-and anti-apoptotic signal transduction pathways determines the viability of a cell when exposed an adverse environmental stress. The radioprotective effect of bFGF against radiation-induced damage to the pulmonary parenchyma confirms that this model also operates *in vivo*. Improved understanding of the molecular ordering and temporal evolution of stress-induced apoptosis may provide opportunities to pharmacologically modulate lethal or mutagenic effects of environmental stresses. The potential for developing approaches to modify the radiation response *in vivo* is particularly attractive due to the limited success of attempts to modulate clinical radiation sensitivity or resistance in human tumors and normal tissues.

References

1 Wyllie AH *Nature*. **284**: 554-556, 1980

2 Raff MC, Barres BA, Burne JF, et al.: *Science*. **262**: 695-700, 1993

3. Haimovitz-Friedman A, Kolesnick RN, Fuks Z: *Sem Rad Oncol*. **6**:273-283, 1996

4. Ward W.F., Sharplin J., Franko AJ, Hinz JM: *Rad Res*. **120**:113-120, 1989

5. Radford IR: *Int J Rad Biol Rel Stud in Phys, Chem Med*. **49**: 611-620, 1986

6. Thompson LH, Suit HD: *Int J Rad Biol Rel Stud in Phys, Chem Med*. **15**: 347-362, 1969

7. Haimovitz-Friedman A, Kolesnick RN, Fuks Z: *British Medical Bulletin*. In Press, 1997

8. Ellis HM, Horvitz HR: *Cell*. **44**: 817-829, 1986

9 Haimovitz-Friedman A, Kan C-C, Ehleiter D.et al.: *J Exp Med*. , **180**: 525-535, 1994

10 Verheij M, Bose R, Lin XH, et al.: *Nature*. **380**: 75-79, 1996

11. Kolesnick R, Golde DW.:*Cell*. **77**:325-328, 1994

12. Spiegel S, Foster D, Kolesnick R: *Curr Opin Cell Biol.* **8**: 159-67, 1996

13. Obeid LM, Linardic CM, Karolak LA, Hannun YA: *Science.* **259**:1769-1771, 1993

14. Jarvis WD, Kolesnick RN, Fornari FA, et al.: *Proc Nat Acad Sci USA.* **91**: 73-77, 1994

15. Santana P, Pena LA, Haimovitz-Friedman A, et al.: *Cell.* **86**:189-199, 1996

16. Chen YR, Wang XP, Templeton D, et al.: *J Biol Chem.* **271**: 31929-31936, 1996

17. Nishina H, Fischer KD, Radvanyi L, et al. *Nature.* , **385**: 350-353.1997

18. Emoto Y, Manome Y, Meinhardt G, et al.: *EMBO J.* **14**: 6148-6156, 1995

19. Korsmeyer SJ, Yin XM, Oltvai ZN, et al.: *Bioch Bioph Acta.* **1271**: 63-66, 1995

20. Gajewski,TF, Thompson CB: *Cell.* **87**: 589-592, 1996

21. Canman CE, Kastan MB: *Sem Canc Biol.* **6**: 17-25, 1995

22. Lowe SW, Schmitt EM, Smith SW, et al.: *Nature.* **362**: 847-849, 1993

23. Beg AA, Baltimore D: *Science.* **274**: 782-784, 1996

24. Wang CY, Mayo MW, Baldwin AS: *Science.* **274**: 784-787.1996

25. Van Antwerp DJ, Martin SJ, Kafri T, et al.: *Science.* **274**: 787-789, 1996

26. Fuks Z, Persaud, RS Alfieri A, et al.: *Canc Res.* **54**: 2582-2590, 1994

27. Haimovitz-Friedman A, Balaban N, McLoughlin M, Ehleiter D, et al.:*Canc Res.* **54**: 2591-2597, 1994

28. Ballou LR, Laulederkind SJ, Rosloniec EF, et al.: *Bioch Bioph Acta.* **1301**:273-287, 1996

29. Jarvis WD, Fornari FA, Browning JL, et al.: *J Biol Chem.* **269**: 31685-31692, 1994

30. Xia ZG, Dickens M, Raingeaud J, et al.: *Science.* **270**: 1326-1331, 1995

31. Hallahan DE, Virudachalam S, Sherman MA, et al.: *Canc Res.* **51**: 4565-4569, 1991

32. Hallahan D, Virudachalam ES, Gardina D, Weichselbaum RR: *Int J Rad Oncol Biol Phys.* **24**::687-692:1992

33. Kim CY, Giaccia AJ, Strulovici B, Brown JM: *Br J Canc.* **66**: 844-849, 1992

34. Uckun FM, Schieven GL, Tuel Ahlgren LM, et al.: *Proc Natl Acad Sci USA.* **90**:252-256, 1993

35. Fuks Z, Alfieri A, Haimovitz-Friedman A, et al.: *Canc J Sci Am.* **1**:62-72, 1995

IV

Radiation-induced DNA Damages, Repair and Mutagenesis

11. Repair of Ionizing Radiation Induced Base Damage: Role of DNA Glycosylases

Serge BOITEUX

Commissariat à l'Energie Atomique
Département de Radiobiologie et Radiopathologie
BP6, 92265-Fontenay aux Roses (France)

DNA DAMAGES PRODUCED BY IONIZING RADIATION

Hydroxyl radicals, e^-_{aq} and H atoms are produced when water is exposed to ionizing radiation. In the presence of oxygen, e^-_{aq} and H atoms react with oxygen and are converted into O_2^-. These chemical species can react with DNA either directly or after metal catalysed transformation in the cell. As many as 100 different DNA modifications have been identified after exposure to ionizing radiation (1,2). These modifications include single-strand and double-strand breaks, base modifications, abasic sites and cross-links (1,2). These lesions are primary sites for radiation-induced cell lethality, mutations and malignant transformation. The double-strand breaks in DNA are particularly important in the lethal effect of ionizing radiation. On the other hand the biological importance of base modifications has not been clearly established. However, several studies show that radiation-induced base damage in DNA has biological consequences. A major product in g-irradiated DNA, the 7,8-dihydro-8-oxoguanine (8-OxoG), is a mutagenic lesion yielding GC-->TA transversions (3). Another radiation-induced base damage, 2,6-diamino-4-hydroxy-5-formamidopyrimidine (Fapy-G) is expected to be a lethal lesion which blocks DNA chain elongation (4). Furthermore, repair activities removing radiation-induced base damage in DNA have been identified in living organisms from bacteria to man.

REPAIR OF RADIATION-INDUCED BASE DAMAGE IN PROKARYOTES

The endonuclease III of E.coli is a monomeric protein of 23-kDa, containing 211 amino acids residues. This protein has a $[4Fe-4S]^{2+}$ iron sulfur cluster with a ligation pattern [Cys-X_6-Cys-X_2-

Cys-X_5-Cys] (2,4). Endonuclease III has a DNA glycosylase activity that excises oxidized pyrimidines such as thymine glycol and 5,6-dihydrothymine (2,4). Endonuclease III is also endowed with an AP lyase activity that nicks DNA at abasic site *via* a b-elimination reaction. Endonuclease III is coded for by the *nth* gene in *E.coli*. An endonuclease III-deficient strain is not unusually sensitive to ionizing radiation or other DNA damaging treatments. The *nth*⁻ mutant, however displays a weak spontaneous mutator phenotype (2,4).

The Fpg protein of E.coli is a monomeric protein of 30-kDa, containing 269 amino acids residues. The Fpg protein has a zinc-finger consensus sequence [Cys-X_2-Cys-X_{16}-Cys-X_2-Cys] which is localised at the carboxy-terminal end of the protein. As the endonuclease III, the Fpg protein is a glycosylase / AP lyase. The Fpg protein excises me-Fapy-G, Fapy-G, Fapy-A and 8-OxoG (4). The *fpg*⁻ mutant exhibits a moderate mutator phenotype. In contrast the double mutant *fpg mut*Y of *Escherichia coli* is a strong spontaneous mutator which specifically accumulates GC-->TA transversions (5,6).

Other bacterial Fpg proteins. Enzymatic activities that excise Fapy-G and/or 8-OxoG have been identified in bacteria such as *M.luteus*, *B.subtilis* or *T.thermophilus*. The *fpg* gene from the Gram-positive bacteria *Lactococcus lactis* has been cloned and the protein purified to homogeneity (6). The *fpg* gene of *L.lactis* is predicted to encode a protein of 273 amino acids with 59% similarity and 38% identity with the *E.coli* Fpg protein. Comparison of physical, catalytic and immunological properties of the Fpg proteins from *E.coli* and *L.lactis* shows that these two proteins have highly conserved sequences and functions. Searches in databases reveals five proteins of bacterial origin displaying strong sequence homologies with Fpg proteins of *E.coli* and *L.lactis*. (Fig.1). These proteins have between 269 and 275 amino acids and have two highly conserved regions. These regions are the [PELPEVET] sequence at the amino terminal which is present in all five proteins and the zinc-finger sequence at the carboxy-terminal [Cys-X_2-Cys-X_{16}-Cys-X_2-Cys-$X_{2\text{-}4}$-COOH] (Fig.1).

E.coli	**PCRVCGTPIVATKHAQRATFYCRQCQK**	**(269)**
L.lactis	**KCSRCGEAIQKIKVAGRGTHFCPVCQQK**	**(273)**
H.influenzae	**PCPKCGGKIESLIIGQRNSFFCPKCQKRG**	**(271)**
N.meningitidis	**PCPQCGGLVVKETLGQRGTFYCPNCQK**	**(275)**
B.firmus	**PCRQCGTDILKTVVGGRGTHFCPNCQL**	**(274)**
Consensus	**PC··CG··I·······R·T··C··CQ·**	

Figure 1: Comparison of the amino acid sequence of five Fpg proteins from bacterial origin. The number of amino acids is indicated and residues chelating zinc atom are underlined.

The MutY protein of E.coli is a monomer of 350 amino acids with a molecular weight of 39-kDa containing a $[4Fe\text{-}4S]^{2+}$ iron sulfur cluster (3,5,7). The MutY protein possesses a DNA glycosylase activity that catalyses the removal of adenine residues incorporated opposite 8-OxoG or guanine. The MutY protein also excises adenine placed opposite cytosine or 8-OxoA. This is an example of a repair enzyme removing a normal DNA base (adenine) placed opposite a lesion (8-OxoG or 8-OxoA) (3,5,7). There has been some controversy to whether the MutY protein has an associated AP nicking activity. Recent studies show that MutY is not a glycosylase / AP lyase, since it is unable to form a covalent intermediate with DNA (8). The *mut*Y⁻ mutant exhibits a strong spontaneous GC-->TA mutator phenotype (5). The MutY protein acts synergistically with the Fpg protein to prevent mutations by 8-OxoG. The Fpg/MutY or GO repair pathway is a major component of cellular defense against the deleterious action of endogenous reactive oxygen species in bacteria (5,9).

REPAIR OF RADIATION-INDUCED BASE DAMAGE IN *SACCHAROMYCES CEREVISIAE*

The Ogg1 protein is a monomeric protein of 43-kDa containing 376 amino acids residues (10). The *OGG1* gene of *S.cerevisiae* which codes for this repair activity was cloned and sequenced (10). The Ogg1 protein, which has been purified to apparent homogeneity is endowed with a glycosylase activity that excises 8-OxoG and me-Fapy-G from DNA. The Ogg1 protein removes 8-OxoG in the 8-OxoG/C duplex but not in the 8-OxoG/A duplex as previously reported for the Fpg protein of *E.coli*. However, the substrate specificities of these two proteins are significantly different. The Fpg protein releases me-Fapy-G and 8-OxoG at the same rate whereas Ogg1 releases 8-OxoG 10 to 20-fold faster than me-Fapy-G (10). The Fpg protein incises DNA at AP site by b- and d-elimination reactions whereas Ogg1 catalyses only the b-elimination reaction. The Ogg1 and Fpg proteins have similar biological functions, since the expression of the *OGG1* gene in *fpg mut*Y mutant of *E.coli* partially suppresses the spontaneous mutator phenotype. The *OGG1* gene is not essential, since the disruption of the *OGG1* gene of *S.cerevisiae* does not affect the viability of resulting haploid *ogg1* mutant (11). The *ogg1* mutant is a spontaneous mutator which specifically accumulates GC-->TA transversions (11). Table 2 shows a comparison of the properties of the Fpg protein of *E.coli* and the Ogg1 protein of *S.cerevisiae*. The similarity of the enzymatic activities and properties of the defective mutants suggest that the Ogg1 protein is the functional homologue of the Fpg protein. However, comparison of the amino acid sequence of the Ogg1 protein with that of the Fpg protein does not reveal an obviously conserved region. Both the highly conserved [PELPEVET] sequence at the amino terminal and the zinc-finger motif at the carboxy-terminal ends of bacterial Fpg proteins are absent in the Ogg1 protein (Fig.1).

	Fpg protein	Ogg1 protein
GENE LEVEL		
Structural gene	*fpg*	*OGG1*
Localisation	82 min	Chromosome XIII
PROTEIN LEVEL		
Size (amino acids)	269	376
Molecular Mass (kDa)	30	43
Metal Ion	Zn	ND*
Zinc-finger motif	Yes	No
Cross-reactivity of antibodies	No	No
ENZYME ACTIVITY LEVEL		
DNA glycosylase	Yes	Yes
Excision of 8-OxoG	+++	+++
Excision of Fapy	+++	+
AP-lyase activity	Yes	Yes
b-elimination	Yes	Yes
d-elimination	Yes	No
MUTANT LEVEL	*fpg::Kan*	*ogg1::TRP1*
Viability	Yes	Yes
Mutator	Yes	Yes
GC-->TA transversions	Yes	Yes
Cross-Complementation	Yes	ND*

ND*: not determined.

Figure 2: Comparison of the properties of the Fpg protein of *E.coli* and Ogg1 protein of *S.cerevisiae*.

The Ogg2 protein is a monomeric protein of about 40-kDa which cleaves DNA at 8-OxoG/G mismatch (12,13). The Ogg2 protein is also a me-Fapy-G DNA glycosylase which does not require cofactor for full enzymatic activity. Cleavage of the 8-OxoG/G duplex probably results from the excision of 8-OxoG residue followed by b-elimination at the resulting AP site (12,13). Therefore, as Ogg1, the Ogg2 protein should be a glycosylase / AP lyase. The Ogg2 protein activity is detectable in cell free extracts of the Ogg1-deficient strains (13,14). Taken together, these observations demonstrate that the Ogg1 and Ogg2 proteins are not products of the same gene. However, cloning of the *OGG2* gene has not been reported yet.

The Ntg1 protein. One locus on chromosome I, *NTG1*, encodes a protein with sequence similarity to the endonuclease III of *E.coli*. The *NTG1* gene codes for a protein of 399 amino acids with of molecular weight of 45-kDa (15). Despite sequence homology with the endonuclease III, the Ntg1 protein does not possess an iron-sulfur DNA binding site. The Ntg1 protein is a DNA glycosylase / AP lyase which excises me-Fapy-G and cleaves DNA containing oxidized pyrimidines. The Ntg1 protein does not cleave DNA duplexes with 8-OxoG placed opposite any one of the four DNA bases (15). The Ntg1-deficient strain is moderately sensitive to oxidizing agents such as H2O2 and menadione. These results strongly suggest that Ntg1 and Ogg2 are not products of the same gene.

REPAIR OF RADIATION-INDUCED BASE DAMAGE IN MAMMALIAN CELLS

Fapy DNA glycosylase activities have been identified in rodent liver and calf thymus cell free extracts (16,17). The calf thymus enzyme has been purified more than 25,000-fold and has an apparent molecular weight of 30-kDa (17). Repair activities for 8-OxoG have also been partially characterised in human cells (18). The results indicate that HeLa cells possess two enzymatic activities that recognize 8-OxoG in DNA: one is a DNA glycosylase that releases 8-OxoG as a free base and the other is an 8-OxoG endonuclease that cleaves DNA at both sides of 8-OxoG. The capacity of these two activities to repair me-Fapy-G residues has not been investigated. Genes coding for these various repair activitie have not been cloned yet.

A Thymine glycol DNA glycosylase activity has been identified in mammalian cells and partially purified. Recently, a DNA glycosylase which excises thymine hydrate and thymine glycol has been purified to apparent homogeneity from calf-thymus (19). This protein of 30-kDa is a DNA glycosylase / AP lyase which forms a covalent intermediate with the DNA substrate in presence of the reducing agents NaBH4 (19). Three peptides were sequenced and searches in database has permitted the identification of a human homologue (19).

The MYH protein is the mammalian homologue of the *E.coli* MutY protein. The calf thymus MYH enzyme was purified 1600-fold. It has an apparent molecular weight of 65-kDa and possesses an activity that cleaves DNA at adenine residues when paired with 8-OxoG (20). The MYH protein cross-reacts with antibodies raised against the MutY protein of *E.coli*. Searches in database has permitted the identification of a human sequence that displays 41% identity to the *E.coli* MutY protein (21). The human homologue of MutY is a 535 amino acid protein that contains an iron sulfur cluster (21). However, the enzymatic properties of this protein have not yet been analysed.

CONCLUSIONS

In prokaryotes as well as in eukaryotes, repair of ionizing radiation-induced base damage in DNA is mostly mediated by the Base Excision Repair pathway (BER). The comparison of the amino acids sequences of the various proteins described in this review suggests two enzyme families: **the Fpg protein family** whose members are only from bacterial origin (Fig.1) and **the endonuclease III family** whose members are from bacterial, archaebacterial and eukaryotic origin. These proteins have a highly conserved sequence that forms a structural motif and contains the glycosylase / AP lyase catalytic site (Fig.2).

	Helix hairpin Helix ——— G/P....D loop
E.coli	**L**EALP**GVG**R**K**T**A**NVVLNT**A**FGW-PTIA**VDTH**IF**R**
H.influenzae	**L**EALA**GVG**R**K**T**A**NVVLNT**A**FGH-PTIA**VDTH**IF**R**
B.subtilis	**L**VKLP**GVG**R**K**T**A**NVVVSV**A**FGV-PAIA**VDTH**VE**R**
M.luteus	**L**VALP**GVG**R**K**T**A**FVVLGN**A**FGQ-PGIT**VDTH**HG**R**
M.jannaschii	**L**LKLP**GVG**R**K**T**A**NLVITL**A**FNK-DGIC**VDTH**VH**R**
S.pombe	**L**MTLP**GVG**P**K**M**G**YLCMSI**A**WNKTVGIG**VDVH**VH**R**
S.cerevisiae[1]	**L**LGLP**GVG**P**K**M**A**YLTLQK**A**WGKIEGIC**VDVH**VD**R**
S.cerevisiae[2]	**L**MSYN**GVG**P**K**V**A**DCVCLM**G**LHMDGIVP**VDVH**VS**R**
C.elegans	**L**CALP**GVG**P**K**M**A**NLVMQI**A**WGECVGIA**VDTH**VH**R**
Consensus	**L GVG K A A VDVH R**

Figure 2: The "endo III" family. The sequence alignment is limited to the HhH motiv and the G/P...D-loop (13).*S.cerevisiae*[1] is the Ntg1 protein and *S.cerevisiae*[2] is the Ogg1 protein.

Today, we cannot exclude the possibility that these two families are related. It is important to notice that all these proteins cleave DNA at abasic sites. This last observation may suggest that all of them have a common ancestor gene coding for an enzyme that recognized abasic sites in DNA.

In bacteria and yeast, the biological function of these DNA glycosylases is to excise lethal and mutagenic lesions induced in DNA by ionizing radiation and chemical oxydants. In mammalian cells, the construction of deficient cell lines and of transgenic animals will provide tools to evaluate the biological role of these repair proteins and their implication in human diseases such as cancer or aging (22,23).

REFERENCES

1.Dizdaroglu,M. Chemical determination of free-radical induced damage to DNA. Free Radical Biol. Med. 1991; 10 : 225-242.

2.Demple,B. and Harrisson,L. Repair of oxidative damage to DNA: Enzymology and Biology. Annu. Rev. Biochem. 1994; 63: 915-948.

3.Grollman,A.P. and Moriya,M. Mutagenesis by 8-Oxoguanine: an ennemy within. Trends in Genet. 1993; 9: 246-249.

4.Boiteux,S. Properties and biological functions of the Nth and Fpg proteins of *Escherichia coli*: two DNA glycosylases that repair oxidative damage in DNA. Photochem. Photobiol. B. 1993; 19: 87-96.

5.Michaels,M.L. and Miller,J.H. The GO system protects organisms from the mutagenic effect of the spontaneous lesion 8-hydroxyguanine (7,8-dihydro-8-oxoguanine). J. Bacteriol. 1992; 174: 6321-6325

6.Duwat,P., De Oliveira,R., Ehrlich,D.S., and Boiteux,S. Repair of oxidative DNA damage in Gram-positive bacteria: the *Lactococcus lactis* Fpg protein. Microbiology.1995; 141: 411-417.

7.Tsai-Wu,J.J., Liu,H.F., and Lu,A.L. *Escherichia coli* MutY has both N-glycosylase and apurinic/apyrimidinic endonuclease activities on A/G and A/C mispairs. Proc. Natl. Acad. Sci. USA. 1992; 89: 8779-8783.

8.Bin Sun, Latham,K.A., Dodson,M.L., and Lloyd,R.S. Studies on the catalytic mechanim of five DNA glycosylases. J. Biol. Chem. 1995; 270: 19501-19508.

9. Boiteux,S. and Laval,J. Repair of Oxidized Purines. In "DNA Base Excision Repair", I.Hickson ed, R.G.Landes Company Austin. 1996, in press.

10.Auffret van der Kemp,P., Thomas,D., Barbey,R., De Oliveira,R. and Boiteux,S. Cloning and expression in *E.coli* of the *OGG1* gene of *S.cerevisiae* which codes for a DNA glycosylase that excises 7,8-dihydro-8-oxoguanine and 2,6-diamino-4-hydroxy-5-N-methylformamidopyrimidine. Proc. Natl. Acad. Sci. USA.1996; 93: 5197-5202.

11.Thomas,D., Scott,A., Barbey,R., Padula,M. and Boiteux,S. Inactivation of the *OGG1* gene increases GC-->TA transversions in *S.cerevisiae*. Mol Gen Genet. 1996: in press.

12.De oliveira,R., Auffret van der Kemp,P., Thomas,D., Barbey,R. and Boiteux,S. Formamidopyrimidine DNA glycosylase in the yeast *Saccharomyces cerevisiae*. Nucleic Acids Res. 1994; 22: 3760-3764.

13. Nash,H.M., Bruner,S., Scharer,O.D. et al. Cloning of a yeast 8-oxoguanine DNA glycosylase reveals the existence of a base-excision DNA-repair protein superfamily. Current Biology. 1996; 6: 968-980.

14.Boiteux,S. et al. 1996. Unpublished results.

15.Eide,L., Pirovano,M., Alseth,I., Berdal,K.G., and Seeberg,E. Base excision of oxidative purine and pyrimidine DNA damage in *S.cerevisiae* by a DNA glycosylase with sequence similarity to endonuclease III of *E.coli*. Proc. Natl. Acad. Sci. USA. 1996; 93: in press.

16.Margisson,G.P., and Pegg,A.E. Enzymatic release of 7-methylguanine from methylated DNA by rodent liver extracts. Proc. Natl. Acad. Sci. USA. 1981; 78: 861-865.

17.Laval,J., O'Connor,T.R., D'Herin,C., Auffret van der Kemp,P. and Boiteux,S. Imidazole ring-opened purines: Occurence and repair in *E.coli* and mammalian cells. In: DNA repair mechanisms and their biological implications in mammalian cells. Eds. Laval,J. and Lambert,M., Plenum Press New-York. 1989: 25-36.

18.Bessho,T., Tano,K., Kasai,H., Ohtsuka,E., and Nishimura,S. Evidence for two DNA repair enzymes for 8-hydroxyguanine in human cells. J. Biol. Chem. 1993; 268: 19416-19421.

19.Hilbert,T.P., Boorstein,R.J., Kung,H.C., et al. Purification of a mammalian homologue of *E.coli* endonuclease III. Biochemistry; 35: 2505-2511.

20.McGoldrick,J.P., Yeng-Chen Yeh, Solomon,M., Essigman,J. and Lu,A.L. Characterization of a human homolog of the *E.coli* MutY repair protein. Mol. Cell. Biol. 1995; 15: 989-996.

21.Slupska,M.M., Baikolov,C., Luther,W.M., Chiang,J-H., Wei,Y-F., and Miller,J.H. Cloning and sequencing a human homolog (*hMYH*) of the *Escherichia coli mutY*gene whose function is required for the repair of oxidative DNA damage. J. Bacteriol. 1996; 178: 3885-3892.

22.Ames,B.N. Dietary carcinogens and anticarcinogens. Oxygen radicals and degenerative diseases. Science. 1983; 221: 1256-1264.

23.Breimer,L.H. Molecular mechanisms of oxygen radical carcinogenesis and mutagenesis, the role of base damage. Mol. Carcinogenesis. 1990; 3: 188-197.

12. cDNA Catalog of Fission Yeast (*Schizosaccharomyces pombe*) and Its Application for Cloning of Mammalian DNA Repair Gene

Mitsuoki MORIMYO*[1], Kazuei MITA[1], Etsuko HONGO[1], Tomoyasu HIGASHI[1], Kimihiko SUGAYA[1], Masahiro AJIMURA[1], Masatake YAMAUCHI[1], Satsuki TSUJI[1], Woo-Yoon PARK[2], Shunichi SASANUMA[1], Junko NOHATA[1], Terumi KIMURA[1], Hiromi INOUE[1] and Yoshie ISHIHARA[1]

[1]Genome Research Group, NIRS, Chiba 263, Japan and [2]Chungbuk National University, Korea

In a study of the structure and function of the housekeeping genes of eukaryotic cells, we chose *Schizosaccharomyces pombe* as the model organism and analyzed expressed sequence tags of *S. pombe*. A cDNA library containing more than 12,000 clones was prepared and we have sequenced them all. By a comparison of their DNA and deduced protein sequences with those in private and public databases, they were classified into two groups of 1800 newly identified clones and 900 similar clones with known genes. Among similar cDNA clones, 20 DNA repair genes were identified: 10 known and 10 new genes. Since the housekeeping genes are considered to be conserved through eucaryotic evolution, we can expect to clone human homologue gene by using the conserved amino acid sequence between fission and budding yeast proteins. We have tried to clone mouse DNA helicase gene by the PCR method using the conserved amino acid sequence of a DNA helicase gene in the *S. pombe* cDNA catalog. This method could be applied to any genes as long as their protein homology is highly conserved. In another approach we have cloned stress-inducible genes. It used *lacZ* as a reporter gene and cloned a *lacZ*-fused gene whose expression was dependent on stresses. By these two strategies, cDNA cloning and stress-inducible gene cloning, we have surveyed many genes related to DNA repair and defense against various stresses and found new genes. The analysis of their structures will prvide clues to finding new functions and elucidating the defense mechanism against stress.

INTRODUCTION

S. pombe is one of the simplest eukaryotes. It has a minimum number of housekeeping genes, 7000, and the smallest genome size, 14 Mb, and cells of *S. pombe* are assumed to be a model of human cells. The proteins of housekeeping genes are considered to be conserved through eukaryotic evolution. Some human genes are normally expressed in cells of *S. pombe* as was first shown by Dr. P. Nurse [1]. It is relatively easy to learn the function of a gene by disrupting it by homologous recombination and by introducing a human homologue. Moreover, a recently developed method, the two-hybrid system [2], can be used to clone directly the gene whose product has interaction with a target protein. Based on these points, we have chosen *S. pombe* as the model organism to learn the structure and function of housekeeping genes of eukaryotes. Of course, DNA repair and stress-response genes are important housekeeping genes. I would like to introduce our trials to attain this goal. They are cDNA catalog of *S. pombe* [3], a cloning trial of mammalian DNA repair gene and a cloning of stress-inducible genes of *S. pombe* [4].

1. cDNA catalog of fission yeast *S. pombe*

The cDNA library of *S. pombe* was prepared by lysing cells with Novozyme, extracting

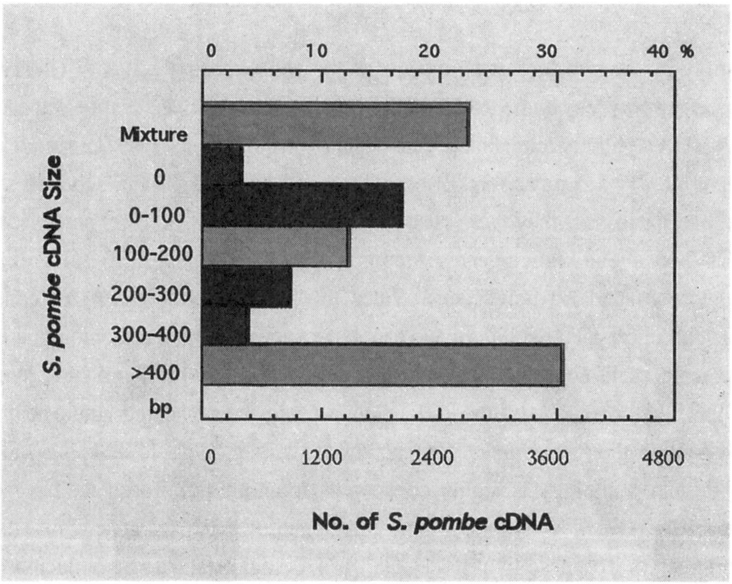

Fig. 1 The insert size and population of *S. pombe* cDNA. The cDNA synthesized and cloned in M13 vector was sequenced by cycle-sequencing method using fluorescent dye-primer. 12000 clones were examined and their insert size and population are shown.

polyA+ RNA and cloning cDNA synthesized by reverse transcriptase into M13 sequencing vector. Without amplification, 12000 plaques were obtained by transformation and a single-stranded DNA was extracted for DNA sequencing, because it gave better result than double-stranded DNA, longer reading and higher quality. Even using ss-DNA, we could read DNA sequence from both directions by devising a new method[5]. We have sequenced them all and analyzed them by searching for homology in private and public databases. By searching for homology in a private database, cDNA clones were classified into groups and the independent clones which had no homologies with each other were counted to be 2700. Homology search of those independent clones in public databases identified each clone. The summary of cDNA quality such as the insert size and its population is shown in Fig. 1. Among 12000 cDNA clones examined, 30% of cDNA clones had over 400 bp,

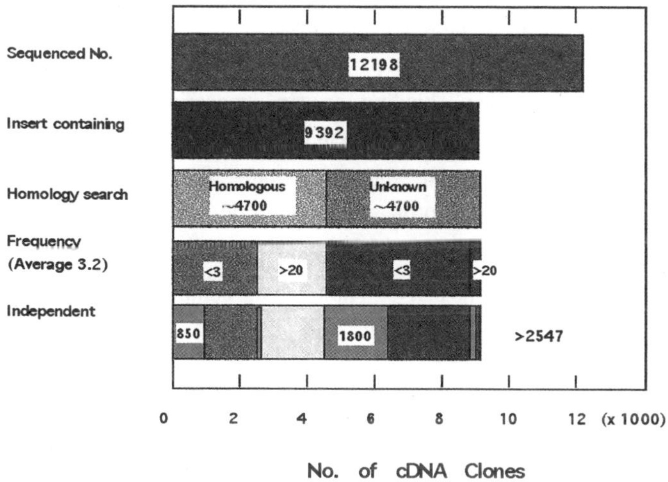

Fig. 2 Analysis of *S. pombe* cDNA by homology search. The cDNA clones were sequenced and their homology with genes deposited in public and private databases was searched by Smith-Waterman algorithm. They were classified into two groups; a homologous group whose clones have homology with known genes and an unknown group whose clones do not have homology with known genes. Independent clones numbered 2800 after searching for homology in a private cDNA database.

30% ranged from 100 bp to 400 bp, 20% carryied no insert DNA and had below 100 bp, and 20% showed mixture, indicating multiple infection. These results indicated that the cDNA library contained relatively short insert DNA but the cDNA quality seemed fairly good because we could not find a clear chimera.

The homology search result in public databases is shown in Fig. 2. Among 12000 cDNA clones examined, 9000 clones had insert DNA and were classified into two groups: a homologous group and an unknown group. The homologous group indicates that its clone is over 30% homologous in amino acid sequence with the gene of other species or identical with the *S. pombe* gene, while the unknown group means that it has no meaningful homology with genes deposited in public databases. Half of homologous clones were highly expressed. That was reasonable because it was easy to purify their proteins and discover their functions. There were about 900 independent clones in the homologous group while there were 1800 in the unknown group. Independent clones numbered, therefore, 2700 as a total and the average frequency of cDNA clone was 3.2. Those independent clones cover over 1/3 of *S. pombe* genes, because they are assumed to number 7000. Homologous clones were classified into groups such as DNA repair, cell cycle, helicase, polymerase, kinase, heat shock protein, binding protein, transcription, translation, ribosomal protein, tRNA synthesis, mitochondria and various kinds of enzymes. We were able to obtain all kinds of genes without bias. The most abundant clone was glyceraldehyde-3-phosphate dehydrogenase gene and it appeared over 100 times among 12000 clones with about 1% frequency. Its product was well known as a major protein, often weighing 20% of the total protein. This is in good accordance with the abundance of the cDNA. Although the complete genome sequence of *S*.

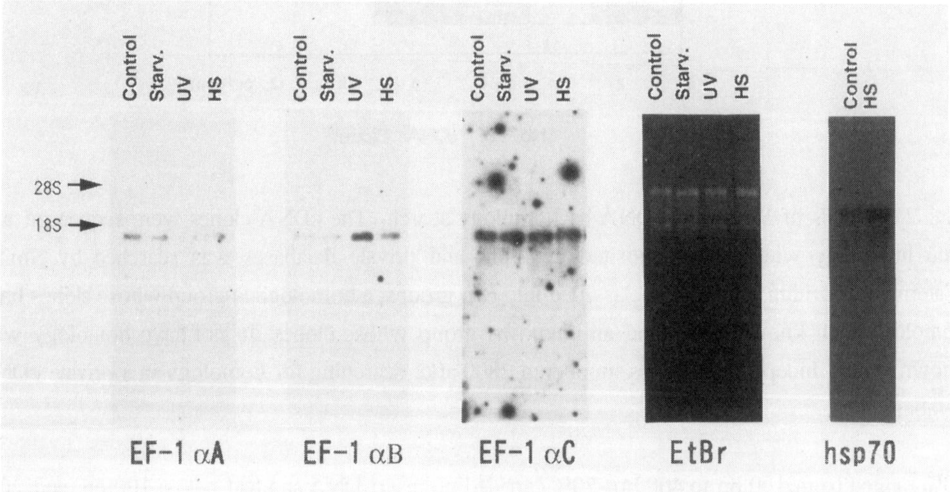

Fig. 3 Northern analysis of influences on transcriptions of EF1-alpha genes with stressful conditions.

cerevisiae was published, the genes are still hypothetical. Our EST data will help identify them clearly.

During this cDNA project, we obtained some unexpected results[6]. One example is shown in Fig. 3. The elongation factor EF1-alpha gene was highly expressed and we obtained more than 50 clones. They could be assembled into a contiguous gene without recloning a complete cDNA clone except the 3'-UTR region. We were aboe to identify three different kinds of 3'-UTRs and clone three isomers of EF1-alpha gene with an almost identical coding region. To learn their functions, we examined the induction pattern after stress-treatments such as starvation, UV and heat shock. We observed distinct UV-induction of EF1-alpha B but not of the others, suggesting that these isomers play a specific role under various conditions.

2. Cloning trial of mammalian DNA repair gene

Among homologous clones, we obatined about 20 cDNA clones related to DNA repair (Table 1). Half of them shown in the closed box were new and 5 clones had no mammalian homologue genes yet. One of the important goals of this cDNA project is to clone mammalian DNA repair genes by yeast cDNA sequences. We discovered a clue to cloning mammalian homologue gene by using the DNA sequence deduced from the amino acid sequence conserved between budding and fission yeast repair proteins, because the housekeeping gene is conserved through eucaryotic evolution. Since these genes have DNA helicase activity, we can expect the conserved amino acid sequences to contain some domain or motif of DNA helicase. We designed PCR primers which were deduced from those conserved 6 to 7 amino acid sequences. After PCR using mouse testis cDNA library as a template, the products were cloned in T-vector and insert DNA sequences were determined. By searching for homologous genes with insert DNA sequences in public databases, we found three clones with DNA helicase activity. A Rad1-12 clone was homologous with *Bacillus* TrcF at 38%/170aa and also with human EST at 98%/132bp. A clone Rad8-2 was homologous with yeast RAD5 at 35%/66aa and human EST at 88%/165bp. And a clone Rad21-35 was homologous with human transcription elongation factor SII at 75%/378bp. The DNA sequencing of 15 cDNA clones obtained from the mouse testis cDNA library by Southern hybridization revealed an ORF of 1182 amino acid protein with 650 bp 5'-UTR and 500 bp 3'-UTR. By designing PCR primers deduced from the amino acid sequence conserved between two kinds of yeast genes, we succeeded in cloning a mouse DNA helicase gene. I believe this strategy is fairly effective for cloning of human homologue as was beautifully shown by the cloning of human *RAD51* gene[7].

3. Cloning of stress-inducible genes of *S. pombe*

In the cDNA project, we have been trying to identify new genes involved in DNA repair and

Table 1. *S. pombe* cDNA clones having homology with known DNA repair genes.

Clone	Homologous gene	Homology (%)	Repair	Species	Properties
Sp cDNA497	rad26	94.6%/56bp	Checkpoint repair	S. pombe	G2 checkpoint repair
Sp cDNA1167	swi10	80.6%/175bp	Rec-repair	S. pombe	Mating type switching, RAD10
Sp cDNA1228	rad8	23.8%/151aa	Error-prone repair	S. pombe	DNA helicase
Sp cDNA1531	rad4	99.3%/135aa	Error-prone repair	S. pombe	XRCC1 homolog
Sp cDNA1750	RAD16	51.6%/31aa	Rec-repair	S. cerevisiae	Helicase, SNF/RAD54 family
Sp cDNA2701	SNM1	40.3%/62aa	Cross-link repair	S. cerevisiae	Zinc-binding domain, Human MK1AA1
Sp cDNA2730	rad24	84.9%/324bp	Checkpoint repair	S. pombe	G2 checkpoint repair
Sp cDNA3247	rad9	97.2%/180bp	Error-prone repair	S. pombe	Cell cycle arrest
Sp cDNA3478	RAD54	39.0%/41aa	Rec-repair	S. cerevisiae	Helicase, SNF/RAD54 family
Sp cDNA3944	rad25	99.2%/126bp	Checkpoint repair	S. pombe	G2 checkpoint repair
Sp cDNA4003	rad26	96.3%/295bp	Checkpoint repair	S. pombe	G2 checkpoint repair
Sp cDNA4427	rad13	72.1%/444bp	Exc-repair	S. pombe	RAD2=XPGC=Endonuclease
Sp cDNA4671	recA	22.4%/87aa	Rec-repair	Rhizobium	RecA
Sp cDNA4914	RAD57	54.0%/50aa	Rec-repair	S. cerevisiae	RAD57
Sp cDNA5108	rhp54	100.0%/376bp	Rec-repair	S. pombe	RAD54 homolog
Sp cDNA5251	RAD10	54.5%/378bp	Exc-repair	S. cerevisiae	ERCC1, RAD10/RAD1=Endonuclease
Sp cDNA5641	RAD23	30.7%/127aa	Exc-repair	S. cerevisiae	RAD23, Human RAD23 homolog
Sp cDNA8143	RAD25	59.6%/260bp	Exc-repair	S. cerevisiae	DNA helicase, ERCC3=XPBC
Sp cDNA10676	RAD1	61.0%/292bp	Exc-repair	S. cerevisiae	UV endonuclease
Sp cDNA10745	swi6	99.0%/507bp	Rec-repair	S. pombe	Mating type switching

defense against various stresses such as radiation, heat or active oxygen. I would like to show another approach to the cloning of stress-inducible genes. The basic strategy to clone stress-inducible genes is using *lacZ* as a reporter gene and finding a *lacZ*-fused gene whose expression depends on stresses. When a stress-inducible gene is connected in frame to *lacZ* gene, its expression is stress-dependent because *lacZ* expression depends on its original gene expression. To clone stress-inducible genes of *S. pombe*, a vector carrying a multi-cloning site upstream of the *lacZ* region, *ars* of *S. pombe*, *URA3* selection marker and *amp* gene was constructed. A fusion DNAlibrary was prepared by inserting 2-4 kb genomic DNA of *S. pombe* at the multi-cloning site and by amplifying in *E. coli*. Yeast URA^+ transformants were replicated and replica plates were treated with various stresses such as UV, X-rays, heat and oxygen. Plates were incubated to see if colonies exhibited blue color as an indication of stress-induction. Cells transformed only by a vector did not show blue color while those by fusion library did. By the first screening, we obtained 0.6% transformants as stress-inducible colonies and they decreased to 71 by the second screening. Plasmid DNA was extracted and its insert DNA size was determined by agarose gel electrophoresis after restriction enzyme digestion. By this test some clones were revealed to have an identical DNA insert. This was confirmed from the nucleotide sequence of insert DNA. DNA sequence upstream of the insertion site was determined by dye-terminator method using M4 fragment complementary to *lacZ* gene as a primer. By comparing DNA sequences, 71 clones were classified into 26 groups. Some clones were abundant and the maximum redundancy was 12, suggesting that stress-inducible genes were non-randomly obtained by this method. By the DNA sequence of insert DNA, we searched its homologue in public and private cDNA databases. A pF52 clone was identified as a heat-inducible tRNA synthetase homologue, supporting the usefulness of this method to clone stress-inducible genes. These results are summarized in Table 2. By this method, 26 independent clones were obtained as stress-inducible candidate genes. The typical stress-inducible genes are shown colored and we already have some corresponding cDNA clones. The determination of the DNA sequence of these cDNA clones and other new genes will provide insights into the structure and function of the stress-inducible genes.

The worldwide trend in biology is to determine the entire genome sequence of a model organism, from *E. coli* to mouse and even human, not excepting fission yeast. We have tried to make a catalog of fission yeast cDNA and identified over one-third of total yeast genes. However, about half of them were unknown genes. Among known genes, we were able to find 20 DNA repair genes which had homology with genes deposited in public databases. Many of them were from *S. cerevisiae*. The complete genome sequence of budding yeast has recently been reported. With the help of this information, the cDNA catalog of fission yeast was revealed particularly useful for the

Table 2. *S. pombe lacZ* -fusion clones inducible with UV, X-rays, heat and oxygen stresses.

Fusion plasmid	Induction by	Homology with	%	Protein or Gene	cDNA
pF1	Heat				
pF2					
pF3	Heat, Oxygen			RNase H, byr1	C99
pF4					
pF5		YM25_SCHPO	100%/62aa	mt mutator urfa	
pF6	Heat	YMC1_SCHPO	100%/72aa	Hypothetical COX1 intron1 gene	C362
pF8				unknown	C2904
pF9					
pF10					
pF11	UV				
pF15					
pF17	UV			unknown	
pF19	X-rays			unknown	C2690
pF27				unknown	C781
pF30	X-rays				
pF31					
pF32	Heat	MISPCOB	90.6%/553bp	cytochrome b 3'-flanking gene	
pF33	Heat	MISPCOB	91.4%/489bp	cytochrome b 3'-flanking gene	
pF34					
pF39		COX1_SCHPO	100%/36aa	cytochrome c oxidase	C10774
pF41	Heat			unknown	
pF43					
pF47					
pF48					
pF49	Heat				
pF52	Heat	SKY2_ECOLI	37.5%/32aa	lys-tRNA synthetase heat-inducible	

cloning of mammalian homologue gene. As a PCR primer, we used the DNA sequence deduced from the amino acid sequence conserved between these two yeast gene products, because the housekeeping genes are considered to be conserved during eukaryotic evolution. Recently it was revealed that positionally cloned genes responsible for various human diseases had high similarity with yeast genes identified from the analysis of the complete yeast genome sequence[8]. These results indicate that the cloning of mammalian homologue gene from the conserved amino acid sequence of two kinds of yeast will shorten the time required for the cloning of relevant genes and facilitate the understanding of human diseases. Fission yeast is a fantastic model organism and its increasing information about cDNA and genomic DNA sequences, in combination with the complete DNA sequence of budding yeast, will accelerate the understanding of new structures and functions of housekeeping genes involving DNA repair and defense against stresses in eukaryotes.

REFERENCES

1. M. G. Lee, P. Nurse: *Nature* **327**: 31-35, 1987.
2. S. Fields, O. Song: *Nature* **340**: 245-6, 1989.
3. M. Morimyo, K. Mita, E. Hongo, M. Yamauchi, S. Tsuji, W. Park, K. Yoneda: *Proceedings of 10th ICRR*, 1995.
4. Morimyo, M., E. Hongo, M. Mita, H. Hama-Inaba, I. Machida: *Proceedings of 34th Annual Meeting of American Society for Cell Biology*, 1994.
5. Mita, K., M. Morimyo, E. Hongo: *Nucleic Acids Res.* **22**, 1507-1508, 1994.
6. Mita, K., M. Morimyo, K. Ito, K. Sugaya, K. Ebihara, E. Hongo, T. Higashi, Y. Hirayama, Y. Nakamura: *Gene* (in press).
7. A. Shinohara, H. Ogawa, Y. Matsuda, N. Ushio, K. Ikeo, T. Ogawa: *Nat Genet* **4**: 239-243, 1993.
8. Oliver, S. G.: *Nature* **379**, 597-600, 1996.

13. DNA Double Strand Break Repair: A Comparison between Mammalian and Yeast Cells

P. A. Jeggo
MRC Cell Mutation Unit, University of Sussex, Brighton, BN1 9RR, UK

Abstract

Analysis of mammalian mutants defective in DNA double strand break (dsb) rejoining has revealed a non-homologous mechanism for the repair of dsbs which utilises the protein complex, DNA-PK. DNA-PK comprises three components, two subunits (Ku70 and Ku80) of a heterodimeric protein termed Ku, and a large catalytic subunit, DNA-PKcs. Yeast homologues of the two Ku subunits have been identified, null mutants constructed, and their analysis has shown that a Ku dependent mechanism of dsb repair exists in yeast, but plays only a minor role in the repair of dsbs. The reduced radiosensitivity of the DNA-PK defective mammalian mutants in S/G2 phases of the cell cycle suggests that mechanism(s) of DNA-PK independent dsb rejoining exist in higher organisms. Homologous recombination is a likely candidate for an alternative mechanism. Taken together the results suggest that dsb repair pathways are conserved between higher and lower eukaryotes but that mammalian cells predominantly rejoin dsbs by a DNA-PK dependent non-homologous process, whereas yeast utilise predominantly a *rad52* dependent homologous recombination process.

Introduction

A DNA dsb is potentially one of the most hazardous lesions a cell is likely to encounter. If unrepaired potential outcomes include chromosome fragmentation and loss of genetic material; if misrepaired translocations and deletions could result. Many external agents, such as ionising radiation create DNA dsbs in cells, either directly or indirectly during repair of other lesions. Dsbs can also arise from free radicals generated during metabolism. Finally, and perhaps surprisingly considering the potential hazardous nature of a dsb, cells create such lesions in their own DNA by enzymatic incision during such essential processes as meiosis, mating type switching and V(D)J recombination. It is therefore not surprising that all cell types examined to date have invested resources into mechanisms for repairing such lesions.

For some time it appeared that higher and lower eukaryotes repair dsbs via fundamentally different mechanisms. Recent studies however, have indicated that the repair mechanisms are essentially conserved between species but that the relative usage of the available mechanisms differs between higher and lower eukaryotes. Here I will discuss recent evidence concerning the mechanisms of dsb rejoining used in mammalian cells, and describe evidence suggesting that at least two distinct pathways of dsb repair exist in mammalian cells.

Mammalian Mutants defective in DNA dsb rejoining

Three complementation groups of rodent mutants (designated IR groups 4, 5, and 7) with defects in DNA dsb rejoining have been described (1). The mutants belonging to these groups are listed in Table 1. The remarkably similar phenotypes of the mutants suggest that they may operate in a single repair pathway. These phenotypes include radiation sensitivity with little or no cross sensitivity to UV, and defects in ability to carry out V(D)J recombination (2-4). Recently, the genes defined by these complementation groups have been identified, and two of their products belong to the same protein complex confirming the suggestions drawn on the basis of the mutant phenotypes (5-9, reviewed in 10).

Table 1 Double Strand Break Repair Defective Rodent Mutants

Complementation Group	Gene Defective	Chromosomal Localisation	Defective cell line
4	XRCC4	5	XR-1
5	XRCC5, *Ku80*	2q33-35	*xrs1-7;* XR-V15B XR-V9B *sexi1,3*
7	XRCC7, *DNA-PKcs*	8	mouse *scid* V-3

The first gene to be identified was XRCC5, whose product proved to be Ku80, the larger subunit of the Ku protein (5, 6, 11). Evidence suggesting this came from two directions. Firstly, the Ku80 gene was reported to map to the same location (2q33-35) as a human gene capable of complementing the group 5 mutants (12). Secondly, the Ku protein had previously been characterised as a ds DNA end-binding protein with properties identical to an end-binding activity missing in the group 5 mutants (11, 13). Partial correction of the defective phenotype was achieved by the introduction of human Ku80 cDNA into group 5 mutants, and more recently, full correction has been obtained by transfection with hamster Ku80 cDNA (5, 6, 14). Finally, a number of group 5 mutants have been shown to harbour mutations in their Ku80 cDNA sequences, and radioresistant revertants regain wild type Ku80 cDNA (14, 15). Taken together these results leave little doubt that Ku80 is the product of XRCC5 and that the Ku protein functions in both dsb rejoining and V(D)J recombination.

To date rodent mutants defective in Ku70, the smaller component of the Ku protein have not been identified but group 6 has been designated for such mutants, in anticipation of their being radiosensitive (1). The relatively high frequency at which Ku80 mutants have been isolated compared with the inability to obtain Ku70 mutants is probably explained by the observation that the CHO-K1 cell line at least, and possibly the V79 cell line, has one functional copy of the *Ku80* gene (*XRCC5*) but two copies of Ku70 (*XRCC6*).

Ku has also been shown to function as the DNA end binding component of DNA dependent protein kinase (DNA-PK) (16, 17). The catalytic subunit of this protein complex is a large protein designated DNA-PKcs. The binding of Ku to DNA ends serves to activate the kinase activity of DNA-PK. Evidence that this association based on 'in vitro' data, represents a functional interaction 'in vivo', is provided by the observation that group 5 mutants also lack DNA-PK activity (18). Interestingly, group 7 mutants, represented by V-3 and *scid* cells, were also shown to lack DNA-PK activity but, in contrast to group 5 mutants, were proficient in DNA end-binding activity (7-9). Additionally, group 7 mutants lack DNA-PKcs by Western blot analysis. This suggested that group 7 mutants might be defective in DNA-PKcs and verification of this includes the observation that yeast artificial chromosomes (YACs) encompassing DNA-PKcs can rescue the defects characteristic of V-3 cells (7). Finally, we and others have recently identified an ochre mutation in the DNA-PKcs of *scid* cells (19, 20). DNA-PKcs is a member of the PI 3-kinase superfamily which include lipid kinases (the original PI 3-kinases) as well as a group of larger proteins termed PI K-related kinases which, where examined, have protein rather than lipid kinase activity (21). All members of the PI 3-kinase superfamily share a conserved kinase domain, but the PI K-related kinases have, in addition, a highly conserved domain at the very carboxy terminus region of the protein (22). The ochre mutation in *scid DNA-PKcs* lies between the kinase domain and the conserved domain specific to the PI K related kinase sub-family, and will therefore result in loss of this conserved region. Loss of this region is likely to result in abolition of the kinase activity, and therefore this mutational change represents a likely candidate for the causative mutation, although verification of this will require sequencing of the entire DNA-PKcs cDNA in *scid* cells.

Recently, the gene defective in group 4 (XRCC4) has also been cloned and shown to encode a 38 kDa protein (23). XRCC4 has no identity to any previously characterised genes, and the protein does not have an identified function at present.

The combined defects of group 4, 5 and 7 mutants in dsb rejoining and V(D)J recombination suggests that the gene products defined by these groups operate in a single repair pathway, and the cloning of the defective genes identifies DNA-PK as playing an essential role in this pathway.

Mechanisms of DNA dsb repair

Combined studies with yeast and higher organisms have delineated three general mechanisms of DNA dsb rejoining, namely homologous recombination, single-strand annealing and non-homologous or illegitimate rejoining (24). This is only a first level classification however, and it is clear that within these categories the mechanisms can be still further separated either genetically and/or mechanistically.

A wealth of evidence has shown that homologous recombination is the major mechanism of dsb rejoining in *S. cerevisiae.* The greatly enhanced sensitivity of haploid compared to diploid yeast cells attests to the fact that homologous recombination provides an efficient, high fidelity mechanism for the repair of DNA dsbs since any coding information lost at the site of the break is restored using the undamaged homologue. In *S. cerevisiae*, homologous recombination is dependent upon *RAD50-58,* and *rad51*, *52* and *54* mutants particularly are highly radiosensitive and defective in DNA dsb rejoining (25). Yeast homologues of Ku70 and 80 have been identified, null mutants constructed and in a wild type background they do not confer radiosensitivity (26-30). These results show that in yeast Ku does not participate in the major, *RAD52* dependent mechanism of dsb repair and moreover, that any repair mechanism involving Ku plays only a minor role in repairing dsbs. *Rad52Ku70* or *80* double mutants do, however, show slightly elevated radiosensitivity compared to *rad52* single mutants, demonstrating that a Ku dependent repair mechanism does function in yeast, but represents a minor pathway (see below) (27-29). Studies involving targetted integration have shown that somatic mammalian cells can undergo homologous recombination, although they do so at lower frequencies compared to yeast cells, and recently mammalian homologues of some of the yeast rad genes have been identified including *RAD51*, *52* and *54* (31, 32).

Single strand annealing (SSA) is a mechanism of dsb rejoining which can occur at short direct repeat sequences in the vicinity of the break. Since the intervening sequences are lost this is inherently a low fidelity rejoining mechanism (33, 34). The genetic requirements for SSA include *RAD1* and

RAD10, and *RAD52* but not other members of the RAD52 epistasis group (35, 36). An analysis of *hprt* mutations arising after irradiation in mammalian cells has shown that breaks can also be rejoined at direct repeat sequences in higher organisms (37).

In contrast to *S. cerevisiae*, mammalian cells have abundant direct DNA-end joining activity, and circumstantial evidence suggests that the majority of breaks in mammalian cells are rejoined by such a mechanism (38-40). In line with this, the integration of exogenous DNA into mammalian chromosomes occurs predominantly at non-homologous sequences, in sharp contrast to yeast cells where integration has an almost obligatory requirement for homology. Recent studies have shown that yeast cells also possess mechanisms of direct DNA end-joining, which include direct abutting of DNA ends, fill in of single strand regions at cohesive ended junctions and the use of microhomology, although these events are rare and it is generally necessary to use a *rad52* background to see them (41-43). Studies with the Ku defective yeast mutants are being carried out and initial studies have shown that Ku plays a role in at least some mechanisms of non-homologous recombination in yeast (28-30).

Mechanism of DNA-PK dependent dsb rejoining in mammalian cells

Three studies have investigated homologous recombination in the group 5 mutants, but to my knowledge such studies have not been carried out for group 4 and 7 mutants. Two studies used non-overlapping deletion plasmids to examine extrachromosomal homologous recombination; one study observed a four fold decrease in recombination in the *xrs* mutants whilst the other study found parental levels (44, 45). A more recent study used a construct bearing two overlapping fragments of the neo gene inserted into chromosomal DNA, so that intra-chromosomal homologous recombination could be examined (46). The recognition sequence for the rare-cutting endonuclease, *I-Sce1,* was inserted into one of these fragments to enable a dsb to be introduced selectively into this region of the chromosome. This was achieved by transient transfection with a plasmid encoding the *I-Sce1* endonuclease. Equal levels of homologous recombination were observed between parental and *xrs* cells suggesting that homologous recombination is not a Ku-dependent process.

In contrast however, a system designed to examine rejoining at regions of microhomology involving short direct repeat sequences flanking an *I-Sce1* site revealed a dramatic decrease in this type of end-joining in the group 5 *xrs-6* cells (47). In addition, the group 5 mutants show a decreased ability to integrate exogenous DNA, a process which is known to occur by non-homologous recombination in mammalian cells (45, 48). These combined results suggest that in mammalian cells Ku, together with DNA-PKcs, play a role in a mechanism of illegitimate end-joining. This suggests that the mechanisms operating in lower and higher eukaryotes are the same but that the percent of breaks rejoined by each mechanism differs between the organisms (fig 1).

Evidence for other mechanisms of dsb rejoining in mammalian cells

The above and additional studies demonstrate that a DNA-PK dependent mechanism of dsb rejoining plays an important role in rejoining breaks in mammalian cells, However, an important feature of group 4, 5 and 7 mutants is that they all show decreased sensitivity to ionising radiation in S or late S/G2 phase of the cell cycle (2, 49, 50). XR-1 cells show an especially dramatic response with S phase cells being nearly as radioresistant as parental cells (50). Indeed, this feature also results in a pronounced tail to the survival of synchronous XR-1 cells, which can be eliminated by "tritium suicide", a technique which results in specific loss of S phase cells following pulse exposure to tritriated thymidine due to the incorporation of tritium, and the subsequent production of dsbs uniquely into cells synthesising DNA. The cell cycle alteration in radiation sensitivity has been shown to be accompanied by a concomitant change in the level of dsb rejoining in both XR-1 and *xrs* cells (50, 51). Wild type hamster cells also show enhanced resistance in S phase although the magnitude of the effect is less pronounced. Since XR-1 cells have recently been shown to be deleted for the XRCC4 gene (23), and *xrs-6* cells appear to have no detectable residual wild type Ku80 transcript, these results indicate the presence of a second pathway of dsb rejoining which plays a major role in S or late S/G2 (15). Homologous recombination clearly represents a likely candidate for such a mechanism. The corollary to this is that the DNA-PK dependent mechanism of dsb rejoining plays its major role during the G1 phase of the cell cycle.

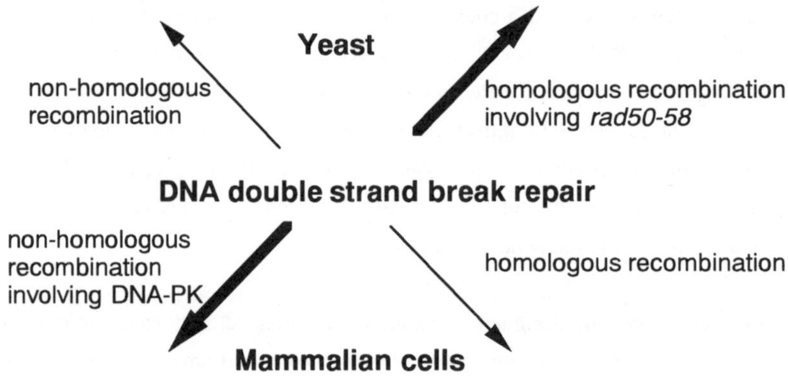

Figure 1. Relative usage of double strand break repair mechanisms in yeast and mammalian cells.

Conclusions

Recent advances in the field of DNA repair have resulted in the identification of mammalian homologues of a variety of DNA repair genes characterised from lower eukaryotes and prokaryotes (reviewed in (25). These and additional mechanistic studies have led to the general conclusion that DNA repair pathways, including nucleotide excision repair (NER), base excision repair (BER), repair of alkylation damage and mismatch correction, are highly conserved between organisms although the precise details of the processes may differ. In contrast, the evidence available for some time has indicated that dsbs are repaired by different mechanisms in mammalian cells compared to lower eukaryotes. The data discussed here suggest that, in fact, the repair pathways are conserved between organisms, but rather that the preferred pathway utilised differs. *S. cerevisiae* repairs the majority of its breaks by homologous recombination, whereas mammalian cells utilise predominantly a non-homologous or illegitimate recombination process. This is perhaps surprising considering the potential lethality of an unrepaired dsb. Homologous recombination provides a dsb repair mechanism of high fidelity since any information lost at the site of the break can be recovered from the undamaged homologue. The use of direct endjoining is likely to enhance the risk of loss of sequences at the rejoined junction, particulalry at sites of complex breaks involving base damage, which arise after radiation damage. If mammalian cells choose to use a lower fidelity mechanism to repair their breaks then there must be a selective advantage to do so. One possibility is that the abundance of repetitive sequences in the mammalian genome may result in homologous recombination being a low, rather than a high, fidelity repair mechanism, if recombination can occur between random repeat sequences. Recent studies have indeed shown that homologous recombination can occur between *alu* sequences in yeast cells harbouring YACs (52). Alternatively, it is possible that infrequent loss of bases at a break site is of less consequence to a mammalian cell carrying a large amount of redundant sequences. These possibilities are not mutually exclusive however, and may have operated together to drive the change. It is also significant that both lower eukaryotes and mammalian cells employ a dsb as an intermediate in a developmental mechanism (mating type switching in yeast and V(D)J recombination in higher organisms), and that these processes occur by the dsb rejoining mechanism predominantly utilised in that organism. Thus mating type switching represents a mechanism of gene conversion, whilst V((D)J recombination occurs by a non-homologous rejoining process. The study of organisms evolutionary intermediate between yeast and man should reveal clues to the significance of these differences.

Acknowledgements

I thank members of my own laboratory, and the laboratories of Dr. S. P. Jackson and G. E. Taccioli, for their contributions to the work discussed here. I apologise for references omitted due to lack of space. I thank Dr. A. R. Lehmann for his varied contributions to this work.

References

1. Thompson, L.H., Jeggo, P.A. *Mutation Res.* 1995, 337:131-134.
2. Jeggo, P.A. *Mutation Res.* 1990, 239:1-16.

3. Taccioli, G.E., Rathbun, G., Oltz, E., Stamato, T., Jeggo, P.A., Alt, F.W. *Science*. 1993, 260:207-210.

4. Pergola, F., Zdzienicka, M.Z., Lieber, M.R. *Molec. Cell. Biol.* 1993, 13:3464-3471.

5. Taccioli, T.G., Gottlieb, T.M., Blunt, T., Priestley, A., Demengeot, J., Mizuta, R., Lehmann, A.R., Alt, F.W., Jackson, S.P., Jeggo, P.A. *Science*. 1994, 265:1442-1445.

6. Smider, V., Rathmell, W.K., Lieber, M.R., Chu, G. *Science*. 1994, 266:288-291.

7. Blunt, T., Finnie, N.J., Taccioli, G.E., Smith, G.C.M., Demengeot, J., Gottlieb, T.M., Mizuta, R., Varghese, A.J., Alt, F.W., Jeggo, P.A., Jackson, S.P. *Cell*. 1995, 80:813-823.

8. Kirchgessner, C.U., Patil, C.K., Evans, J.W., Cuomo, C.A., Fried, L.M., Carter, T., Oettinger, M.A., Brown, J.M., Iliakis, G., Mehta, R., Jackson, M. *Science*. 1995, 267:1178-1183.

9. Peterson, S.R., Kurimasa, A., Oshimura, M., Dynan, W.S., Bradbury, E.M., Chen, D.J. *Proc. Natl. Acad. Sci. USA*. 1995, 92:3171-3174.

10. Jeggo, A.P., Taccioli, G.E., Jackson, S.P. *Bioessays*. 1995, 17:949-957.

11. Getts, R.C., Stamato, T.D. *J. Biol. Chem.* 1994, 269:15981-15984.

12. Hafezparast, M., Kaur, G.P., Zdzienicka, M., Athwal, R.S., Lehmann, A.R., Jeggo, P.A. *Somatic Cell Molec. Genet.* 1993, 19:413-421.

13. Rathmell, W.K., Chu, G. *Proc. Natl. Acad. Sci. USA*. 1994, 91:7623-7627.

14. Errami, A., Smider, V., Rathmeli, W.K., He, D.M., Hendrickson, E.A., Zdzienicka, M.Z., Chu, G. *Molec. Cell. Biol.* 1996, 16:1519-1526.

15. Singleton, B.K., Priestley, A., Gell, D., Blunt, T., Jackson, S.P., Lehmann, A.R., Jeggo, P.A. 1996, submitted.

16. Dvir, A., Peterson, S.R., Knuth, M.W., Lu, H., Dynan, W.S. *Proc. Natl. Acad. Sci. USA*. 1992, 89:11920-11924.

17. Gottlieb, T.M., Jackson, S.P. *Cell*. 1993, 72:131-142.

18. Finnie, N.J., Gottlieb, T.M., Blunt, T., Jeggo, P.A., Jackson, S.P. *Proc. Natl. Acad. Sci. USA*. 1995, 92:320-324.

19. Blunt, T., Gell, D., Fox, M., Taccioli, G.E., Jackson, S.P., Lehmann, A.R., Jeggo, P.A. *Proc. Natl. Acad. Sci. USA*. 1996, 93:10285-10290.

20. Danska, J.S., Holland, D.P., Mariathasan, S., Williams, K.M., Guidos, C.J. *Molec. Cell. Biol.* 1996, 16:5507-5517.

21. Hartley, K.O., Gell, D., Smith, G.C.M., Zhang, H., Divecha, N., Connelly, M.A., Admon, A., Lees-Miller, S.P., Anderson, C.W., Jackson, S.P. *Cell*. 1995, 82:849-856.

22. Kapeller, R., Cantley, L.C. *BioEssays*. 1994, 16:565-576.

23. Li, Z., Otevrel, T., Gao, Y., Cheng, H.-L., Seed, B., Stamato, T.D., Taccioli, G.E., Alt, F.W. *Cell*. 1995, 83:1079-1089.

24. Haber, J.E. *BioEssays*. 1995, 17:609-620.

25. Friedberg, E.C., Walker, G.C., Siede, W. *ASM Press, Washington, USA*. 1995,

26. Feldmann, H., Winnacker, E.L. *J. Biol. Chem.* 1993, 268:12895-12900.
27. Siede, W., Friedl, A.A., Dianova, I., Eckardt-Schupp, F., Friedberg, E.C. *Genetics*. 1996, 142:91-102.
28. Milne, G.T., Jin, S., Shannon, K.B., Weaver, D.T. *Molec. Cell. Biol.* 1996, 16:4189-4198.
29. Boulton, S.J., Jackson, S.P. *EMBO J.* 1996, 15:5093-5103.
30. Boulton, S.J., Jackson, S.P. *Nucl. Acids Res.* 1996, in press.
31. Shinohara, A., Ogawa, H., Ogawa, T. *Cell.* 1992, 69:457-470.
32. Muris, D.F.R., Bezzubova, O., Buerstedde, J.M., Vreeken, K., Balajee, A.S., Osgood, C.J., Troelstra, C., Hoeijmakers, J.H.J., Ostermann, K., Schmidt, H., Natarajan, A.T., Eeken, J., Lohman, P., Pastink, A. *Mutation Res.* 1994, 315:295-305.
33. Lin, F.-L., Sperle, K., Sternberg, N. *Molec. Cell. Biol.* 1990, 10:103-112.
34. Fishman-Lobell, J., Rudin, N., Haber, J. *Molec. Cell. Biol.* 1992, 12:1292-1303.
35. Fishman-Lobell, J., Haber, J.E. *Science*. 1992, 258:480-484.
36. Ivanov, E.L., Sugawara, N., Fishman Lobell, J., Haber, J.E. *Genetics*. 1996, 142:693-704.
37. Morris, T., Thacker, J. *Proc. Natl. Acad. Sci. USA*. 1993, 90:1392-1396.
38. Roth, D.B., Wilson, J.H. *Proc. Natl. Acad. Sci. USA*. 1985, 82:3355-3359.
39. Roth, D.B., Wilson, J.H. *Molec. Cell. Biol.* 1986, 6:4295-4304.
40. Roth, D.B., Porter, T.N., Wilson, J.H. *Molec. Cell. Biol.* 1985, 5:2599-2607.
41. Kramer, K.M., Brock, J.A., Bloom, K., Moore, J.K., Haber, J.E. *Molec. Cell. Biol.* 1994, 14:1293-1301.
42. Mezard, C., Nicolas, A. *Molec. Cell. Biol.* 1994, 14:1278-1292.
43. Moore, J.K., Haber, J.E. *Molec. Cell. Biol.* 1996, 16:2164-2173.
44. Moore, P.D., Song, K.-Y., Chekuri, L., Wallace, L., Kucherlapati, R.S. *Mutation Res.* 1986, 160:149-155.
45. Hamilton, A.A., Thacker, J. *Molec. Cell. Biol.* 1987, 7:1409-1414.
46. Liang, F., Romanienko, P., Weaver, D., Jeggo, P.A., Jasin, M. *Proc. Natl. Acad. Sci. USA*. 1996, in press.
47. Liang, F., Jasin, M. *J. Biol. Chem.* 1996, 271:14405-14411.
48. Jeggo, P.A., Smith-Ravin, J. *Mutation Res.* 1989, 218:75-86.
49. Whitmore, G.F., Varghese, A.J., Gulyas, S. *Int. J. Radiat. Biol.* 1989, 56:657-665.
50. Giaccia, A., Weinstein, R., Hu, J., Stamato, T.D. *Somatic Cell Molec. Genet.* 1985, 11:485-491.
51. Mateos, S., Slijepcevic, P., MacLeod, R.A.F., Bryant, P.E. *Mutation Res.* 1994, 315:181-187.
52. Bennett, C.B., Westmoreland, T.J., Snipe, J.R., Resnick, M.A. *Molec. Cell. Biol.* 1996, 16:4414-4425.

14. Genome Analysis of the *AT* Region of Human Chromosome 11 and Identification of a New Gene, *NPAT*, Physically Linked to the *ATM* Gene

Takashi IMAI[1], Masatake YAMAUCHI[1], Naohiko SEKI[2], Takehiko SUGAWARA[1], Masashi SAGARA[1], Toshiyuki SAITO[1], Yoichi MATSUDA[1], Hiroko ITO[1], Takahiro NAGASE[2], Nobuo NOMURA[2] and Tada-aki HORI[1]

[1]Genome Research Group, National Institute of Radiological Sciences, Chiba 263 Japan

[2]Laboratory of Gene Structure I, Kazusa DNA Research Institute, Chiba 292 Japan

Ataxia telangiectasia (*AT*) is an autosomal recessive disease of unknown etiology associated with cerebellar ataxia, oculocutaneous telangiectasia, immunodeficiency, and hypersensitivity to ionizing radiation. Recent linkage analyses localized the major *AT* locus to a sequence of approximately 850 kb between the markers *D11S1818* and *D11S1819* on chromosome 11q22-23. We isolated a total of 20 yeast artificial chromosomes (YAC) clones from three independent libraries, using sequence tagged sites mapped in the *AT* region as primers for PCR-based YAC screening. Further PCR assay for the presence or absence of 16 different DNA markers allowed us to construct and order four YAC contigs at the *AT* region. One of the contigs, which consists of 10 YACs, covers about 2 Mb of DNA at the boundary between 11q22.3 and 11q23.1 and includes the entire region of the major *AT* locus between *D11S11818* and *D11S1819*. Recently, the gene responsible for *AT*, *ATM*, has been identified by positional cloning on the major *AT* locus. The *ATM* was found to be mutated in *AT* patients from all complementation groups, indicating that it is probably the sole gene responsible for this disorder. No experimental data, however, have been reported regarding whether the *ATM* gene can compensate the cellular defects in two *AT* phenotypic hallmarks, radiosensitivity and radioresistant DNA synthesis. Our attempt was to screen more genes transcribed from the major *AT* region in a YAC contig, in order to examine their possible function(s) related to some complicated phenotypes of *AT*. We identified a new gene, designated *NPAT*, from the major *AT* locus. The gene encoded a 1,427-amino acid protein containing nuclear localization signals and phosphorylation target sites by cyclin-dependent protein kinases associated with E2F. The messenger RNA of *NPAT* was detected in all human tissues examined, and its genomic sequence was strongly conserved through eukaryotes, suggesting that the *NPAT* gene may be essential for cell maintenance and may be a member of the housekeeping genes. Analysis of the genomic region of *NPAT* surprisingly revealed that the gene existed only 0.5 kb away from the 5' end of the *ATM* transcript with opposite transcriptional direction. It is possible that the promoter region could be shared by both housekeeping genes and that each gene could influence the expression of the other.

Introduction

Ataxia telangiectasia (AT) is an autosomal recessive gene disorder defined by a wide spectrum of defects involving a severe neurological disorder due to progressive degeneration of the cerebellum, particularly the Purkinje cells, dilation of blood vessels in the eyes, ears and face, immunological defects such as atrophy of the thymus and diminished T cell response, a greatly elevated incidence of tumors, especially of leukemia and lymphomas, and premature aging.[1,2) In addition, *AT* heterozygotes, estimated in approximately 1% of the population, have a five-fold increased risk of neoplasia, particularly breast carcinoma in women,[3) although the disease is known to be an autosomal recessive inheritance.

At the cellular level, *AT* cells exhibit an unusual response to DNA synthesis following ionizing radiation exposure, i.e., radioresistant DNA synthesis.[4,5) There were defects in the checkpoint function at all phases of the cell cycle shortly after irradiation, allowing the cells to continue cycling despite extensive DNA damage.[6,7). *AT* cells also show a delayed induction of proteins associated with cell-cycle arrest, p53, or DNA replication and repair, replication protein A.[7-11) These data indicate that the wild type of *AT* gene product(s) may act in an early step in a specific DNA damage-signaling pathway.[7,10)

The property of radioresistant DNA synthesis was used to perform genetic complementation tests, suggesting that the *AT* could be divided into four complementation groups, AT-A, -C, -D, and -E.[12) Genetic linkage analyses, however, showed that the gene(s) responsible for at least groups A, C, and E was mapped to the same region on chromosome 11q22-q23.[13,15) Although genetic linkage evidence did not support the hypothesis of genetic heterogeneity, the possibility of a cluster of genes at the 11q22-q23 locus could not be ruled out.[16)

Recent haplotype analysis by an international consortium has indicated that the major *AT* locus, *AT-A/AT-C*, is localized to an ~850 kb interval flanked by *D11S1818* and *D11S1819*.[17) More recently, the gene responsible for *AT*, *ATM*, has been identified by positional cloning on the major *AT* locus.[18,19) The *ATM* was found to be mutated in *AT* patients from all complementation groups, indicating that it is probably the sole gene responsible for this disorder.[18) The predicted protein of 3056 amino acids revealed significant sequence similarities to the proteins containing phosphatidylinositol-3 kinases (PI-3 kinases), such as the *Saccharomyces* genes *TOR1*[20) and *TOR2*,[21) the *Scizosaccharomyces pombe rad3* gene,[19) *Saccharomyces ESR1/MEC1*,[22,23) DNA dependent protein kinase,[24) and *Saccharomyces TEL1*.[25,26) It is very interesting that these candidates of the *ATM* family are shown to be involved in mitogenic signal transduction, meiotic recombination, and cell-cycle control.

No experimental data, however, have been reported regarding whether the *ATM* gene can compensate the cellular defects in two *AT* phenotypic hallmarks, radiosensitivity and radioresistant DNA synthesis. It also remains to be determined whether the *ATM* product has PI-3 kinase or DNA-dependent protein kinase activity and/or whether it can bind DNA.

Our attempt was to screen more genes transcribed from the major *AT* region in a YAC contig,[27] in order to examine their possible function(s) related to some complicated phenotypes of *AT*. Here we describe a new housekeeping gene which lies only 0.5 kb apart from the 5' end of the *ATM* transcribed region with the opposite transcriptional direction.

Materials And Methods

Isolation of YACs, cosmid and cDNA clones has been described previously.[27-30] Human multiple tissue Northern blots and zoo blot were purchased from Clontech Laboratories. Hybridization to the Northern blots and the following washing were performed according to the procedure recommended by the manufacturer. The zoo blot was hybridized at 65 °C in 1 M NaCl, 10% dextran sulphate, 0.5% SDS, and 100 μg yeast tRNA per ml. The filter was washed at 55 °C in 2X SSC, 0.1% SDS and subjected to autoradiography.

Results And Discussion

Isolation of cDNA Clones

The major *AT* region on chromosome 11 lies between *D11S1818* and *D11S1819* and is estimated to encompass 850 kb.[17] From a YAC contig, which covered approximately 2 Mb of DNA at the boundary between the Giemsa-positive band 11q22.3 and the Giemsa-negative band 11q23.1 and included the entire region of the major *AT* locus (Table 1 and Fig.1), three YAC clones were chosen and subjected to the island rescue PCR (IRP)[31] to search for the transcribed sequences. Amplified DNA fragments by IRP extending from *Bss* HII or *Sac* II sites to the nearest *Alu* repeat in the YAC clones were used to screen a human testis cDNA library. Three cDNA clones were isolated from the library consisting of $6X10^5$ clones and these, designated as T4-41, T4-51, and T4-92, were confirmed to be derived from at least one of the YACs by Southern blot analysis of the restriction enzyme digested fragments of the YACs. Nucleotide sequence analysis of these clones revealed that no identical nucleotide sequence, except some ESTs, could be identified in any public databases. Because the Southern hybridization analysis also indicated that the T4-41 clone is located at the center of the major *AT* locus, we focused on T4-41 for further characterization.

Northern blot analysis using poly A+ RNA from various human tissues with the T4-41 clone as the probe showed two hybridizing signals of 6.2 and 5.4 kb. These bands were detected in all of the tissues examined, indicating that the gene was widely expressed but the expression levels and the

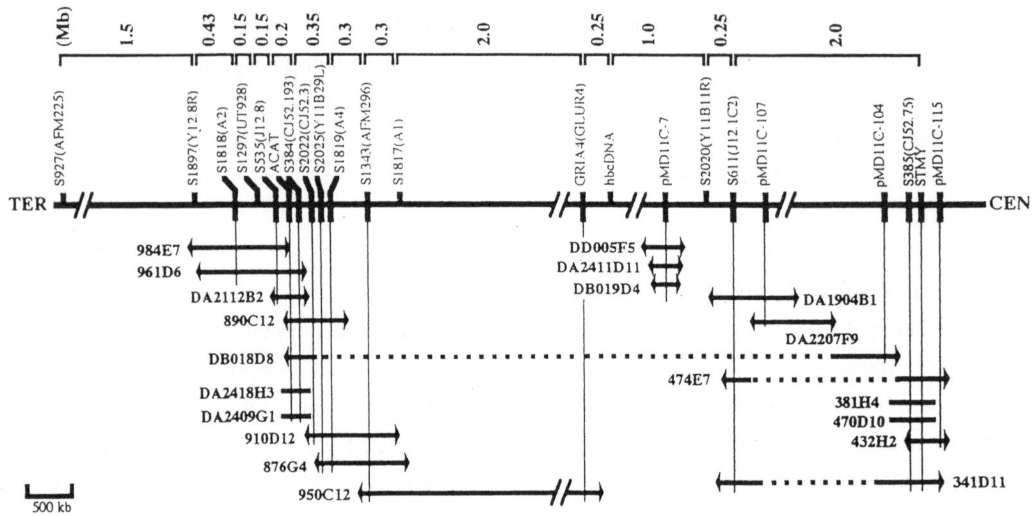

Fig. 1 A YAC-based STS-content map of the *AT* major region.[27] Size and origin of the YACs are summarized in Table 1. STSs used for analysis of the YACs are indicated by a vertical thick bar crossing the horizontal line. The presence of an STS sequence in a YAC is demonstrated by thin line. The YACs represented by dashed line demonstrate that these have discontinuous segments. // indicates that physical length in the maps reported previously are not confirmed by the present YAC contigs.

quantitative ratio of the two RNA species were slightly different in each tissue . An evolutionary conservation test of the T4-41 sequence using genomic DNA from various vertebrates and budding yeast revealed that the gene was strongly conserved through mammals. Relatively weak but positive signals were detected in chicken and yeast DNAs, showing that the gene was conserved through eukaryotes. These results imply that the T4-41 gene may be a member of the housekeeping genes.

Because the insert of the T4-41 cDNA clone did not extend to the length of the poly A+ RNA detected by the Northern blot hybridization, other cDNA libraries from human testis and placenta were further screened using various restriction fragments excised from the insert DNA of T4-41, and four clones positive for hybridization were recovered. Nucleotide sequence analysis of all isolated clones permitted the construction of a 5,900-base continuous sequence excluding a poly (A) tail.

The composite sequence of B-39, T4-41, T4-81, and T4-82, designated *NPAT* (nuclear protein mapped to the *AT* locus, discussed below), was obtained from both the message and complementary strands. There was complete agreement in the nucleotide sequences obtained from the overlapping

Table 1 YAC clones subjected to PCR analyses

DNA marker	YAC clone	Source	Size (kb)
pMD11C-7	DA2411D11	Washington D	360
	DD005F5	Washington D	530
	DB019D4	Washington D	200
pMD11C-104	DB018D8	Washington D	800
	341D11	CEPH	180
pMD11C-107	DA1904B1	Washington D	520
	DA2207F9	Washington D	500/300
pMD11C-115	432H2	CEPH	190
	474 E7	CEPH	330
STMY	381H4	CEPH	n.d.
	470D10	CEPH	n.d.
CJ193	DA2418H3	Washington D	n.d.
(D11S384)	DA2409G1	Washington D	n.d.
	DA2112B2	Washington D	400
	890C12	CEPH	670
	961D6	CEPH	1500
J12.8	961D6	CEPH	1500
(D11S535)	984 E7	CEPH	1200
AFM296	876G4	CEPH	1150
(D11S1343)	910D12	CEPH	1150
	950C12	CEPH	1700

n.d., not determined.

areas. The*NPAT* cDNA contained 66 nucleotides in the 5' noncoding region, 4,281 nucleotides in the coding region, and 1,553 nucleotides in the 3' untranslated region preceding the poly A tail (DDBJ accession number D83243). There is an in-frame stop codon 43 bp upstream of the putative start codon. The 3' untranslated region contains the hexanucleotide 5'-AATAAA-3' which could function as a signal for poly (A) addition or termination of transcription in the expected position, 23 nucleotides upstream from the poly (A) tail.

Predicted Amino Acid Sequence of *NPAT*

The total length of the predicted protein was 1,427 amino acids with a calculated molecular weight of 154,300 (modified residues not included)(Fig. 2). The protein was relatively Ser/Thr rich, 21.3% of the total amino acid residues of the protein and was acidic.

To predict localization of the protein in a cell, the deduced amino acid sequence was subjected to

```
1    MLLPSDVARL VLGYLQQENL ISTCQTFILE SSDLKEYAEH CTDEGFIPAC   50
51   LLSLFGKNLT TILNEYVAMK TKETSNNVPA IMSSLWKKLD HTLSQIRSMQ  100
101  SSPRFAGSQR ARTRTGIAEI KRQRKLASQT APASAELLTL PYLSGQFTTP  150
151  PSTGTQVTRP SGQISDPSRS YFVVVNHSQS QDTVTTGEAL NVIPGAQEKK  200
201  AHASLMSPGR RKSESQRKST TLSGPHSTIR NFQDPNAFAV EKQMVIENAR  250
251  EKILSNKSLQ EKLAENINKF LTSDNNIAQV PKQTDNNPTE PETSIDEFLG  300
301  LPSEIHMSEE AIQDILEQTE SDPAFQALFD LFDYGKTKNN KNISQSISSQ  350
351  PMESNPSIVL ADETNLAVKG SFETEESDGQ SGQPAFCTSY QNDDPLNALK  400
401  NSNNHDVLRQ EDQENFSQIS TSIQKKAFKT AVPTEQKCDI DITFESVPNL  450
451  NDFNQRGNSN AECNPHCAEL YTNQMSTETE MAIGIEKNSL SSNVPSESQL  500
501  QPDQPDIPIT SFVSLGCEAN NENLILSGKS SQLLSQDTSL TGKPSKKSQF  550
551  CENSNDTVKL KINFHGSKSS DSSEIHKSKI EINVLEPVMS QLSNCQDNSC  600
601  LQSEILPVSV ESSHLNVSGQ VEIHLGDSLS STKQPSNDSA SVELNHTENE  650
651  AQASKSENSQ EPSSSVKEEN TIFLSLGGNA NCEKVALTPP EGTPVENSHS  700
701  LPPESVCSSV GDSHPESQNT DDKPSSNNSA EIDASNIVSL KVIISDDPFV  750
751  SSDTELTSAV SSINGENLPT IILSSPTKSP TKNAELVKCL SSEETVGAVV  800
801  YAEVGDSASM EQSLLTFKSE DSAVNNTQNE DGIAFSANVT PCVSKDGGYI  850
851  QLMPATSTAF GNSNNILIAT CVTDPTALGT SVSQSNVVVL PGNSAPMTAQ  900
901  PLPPQLQTPP RSNSVFAVNQ AVSPNFSQGS AIIIASPVQP VLQGMVGMIP  950
951  VSVVGQNGNN FSTPPRQVLH MPLTAPVCNR SIPQFPVPRK SQKAQGLRNK 1000
1001 PCIGKQVNNL VDSSGHSVGC HAQKTEVSDK SIATDLGKKS EETTVPFPEE 1050
1051 SIVPAAKPCH RRVLCFDSTT APVANTQGPN HKMVSQNKER NAVSFPNLDS 1100
1101 PNVSSTLKPP SNNAIKREKE KPPLPKILSK SESAISRHTT IRETQSEKKV 1150
1151 SPTEIVLESF HKATANKENE LCSDVERQKN PENSKLSIGQ QNGGLRSEKS 1200
1201 IASLQEMTKK QGTSSNNKNV LSVGTAVKDL KQEQTKSASS LITTEMLQDI 1250
1251 QRHSSVSRLA DSSDLPVPRT PGSGAGEKHK EEPIDIIKAP SSRRFSEDSS 1300
1301 TSKVMVPPVT PDLPACSPAS ETGSENSVNM AAHTLMILSR AAISRTTSAT 1350
1351 PLKDNTQQFR ASSRSTTKKR KIEELDERER NSRPSSKNLT NSSIPMKKKK 1400
1401 IKKKKLPSSF PAGMDVDKFL LSLHYDE                          1427
```

CDK

cAMP- cGMP dependent PK

protein kinase C

nuclear localization

ATP/GTP binding

hsp90

Fig. 2 Amino acid sequence of the deduced translation product encoded by the *NPAT* gene. Putative nuclear localization signals are boldly underlined. Serine (S) and threonine (T) residues that are potentially phosphorylated by cdk4/E2F or cdk6/E2F complexes (Bold box), cAMP- and cGMP-dependent protein kinase (plain Box), and protein kinase C (dashed box) are indicated. Sequences resembling the ATP/GTP binding domain are marked by thin underlining. The upper double-lines reveals the region homologous to the hsp90 family.

PSORT analysis through the network service of the Institute for Molecular and Cellular Biology, Osaka University,[32] suggesting that the protein may be transported into the nucleus because it had sequences matching the nuclear localization signals consisting of four continuous basic amino acids in the carboxyl terminus (Fig. 2). No existence of a signal sequence or a leader peptide, which is thought to be necessary for protein sorting through the vesicular pathway, was proposed by this analysis. Therefore, the gene was tentatively designated *NPAT* (nuclear protein mapped to the *AT* locus).

The database search with the whole *NPAT* nucleotide and deduced amino acid sequences found no significant similarity with any previously known genes. A detailed computer analysis of smaller fragments in the *NPAT* product showed similarities with a large number of known protein motifs, enabling us to identify the region of possible functional importance. Notable were five consensus sequences for phosphorylation by CDK4/E2F and CDK6/E2F complexes (Ser/Thr-pro-X-Lys/Arg),[33] 24 protein kinase C phosphorylation target sites (Ser/Thr-X-Lys/Arg),[34,35] and three cAMP/cGMP-dependent protein kinase phosphorylation target sequences (Lys/Arg- Lys/Arg-X-Ser/Thr)[36] (Fig. 2).

<u>Genomic Structure of the *NPAT* 5'-Region</u>

To localize the *NPAT* gene on a fine physical map, cosmid libraries were constructed from the three YACs used for the gene identification, and the cosmid clones positive for human DNA were utilized for construction of contigs by screening the clone pool with the genes and the DNA markers

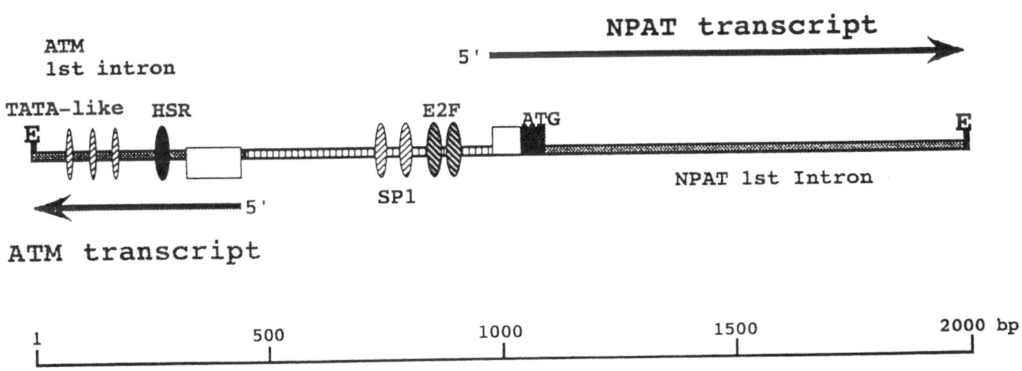

Fig. 3 Schematic diagram of the upstream region nestled between *ATM* and *NPAT* genes. Both genes and transcriptional directions are represented by arrows. The 5' untranslated region in the first exon of both genes is denoted by the white boxes. The protein coding region in the first exon of the *NPAT* gene is marked by a black box with ATG. The first intron of both genes is represented by shaded bars. Possible E2F binding domains, SP1 binding sites, the heat shock responsible element or TATA-like motifs are marked by ellipses. E reveals the *Eco* RI site of the genomic DNA.

which were previously mapped in the major *AT* locus and by cosmid walking. Hybridization of the *NPAT* cDNA with the YAC and the cosmid contigs revealed that the gene was transcribed from telomere to centromere. Through the mapping of restriction enzyme cleavage sites on the cosmid contigs, we found that the same 2 kb-*Eco* RI fragment of the contig was detected with both the oligonucleotide probe corresponding to the 5' of *NPAT* cDNA and the 5'-restriction DNA fragment of the *ATM* gene which is transcribed from centromere to telomere (Fig. 3).[18)]

Nucleotide sequence analysis of the 2 kb-*Eco* RI fragment (DDBJ accession number D83244) showed that these two genes lay 513 bp apart in the opposite transcriptional direction. The database search with the nucleotide sequence revealed that this region contained a 256-bp sequence which was previously predicted to be one of the CpG islands isolated by the methylated DNA binding column.[37)] Three degenerative TATA-like boxes were found in the upstream region at 782 bp, 893 bp, and 918 bp from the 5' end of the *NPAT* cDNA. Also, two GC boxes[38)] existed at 179 and 195 bp upstream of the *NPAT* transcript. Several types of transcriptional regulation motifs including one heat shock responsible element[39)] (102 bp apart from the 5' end of the gene) and two sites of the E2F binding domain[40)] (69 and 113 bp upstream) could be predicted in the 5' region of the *NPAT* gene which partly overlapped the first exon and intron of the *ATM* gene (Fig. 3). It is noteworthy that some of the promoter elements and regulatory sequences work in either orientation. Therefore, this *NPAT* upstream region may regulate the transcription of both genes.

A close physical relationship between the two genes, *ATM* and *NPAT*, in the major *AT* locus may suggest an important mechanism for the initiation and regulation of their expression which could affect the complex phenotypes of *AT*.

References

1. Harnden, D.G., 1994. *Int. J. Radiat. Biol.*, **66**: S13-S19.
2. Shiloh, Y., 1995. *Eur. J. Hum. Genet.*, **3**:116-138.
3. Swift, M., Morrell, D., Massey, R.B., and Chase, C.L., 1991. *New Engl. J. Med.*, **325**: 1831-1836.
4 Houldsworth, J., and Lavin, M.F., 1980. *Nucleic Acids. Res.*, **8**: 3709-3720.
5. Painter, R.B., and Young, B.R., 1980. *Proc. Natl. Acad. Sci. USA*, **77**: 7315-7317.
6. Hartwell, L.H. and Kastan, M.B., 1994. *Science*, **266**: 1821-1828.
7. Thacker, J., 1994. *Int. J. Radiat. Biol.*, **66**: S87-S96.
8. Kastan, M.B., Zhan, Q., El-Deiry, W.S., Carrier, F., Jacks, T., Walsh, W.V., Plunkett, B.S., Vogelstein B., and Fornace, Jr, A.J., 1992. *Cell*, **71**: 587-597.
9. Khanna, K.K. and Lavin, M.F., 1993. *Oncogene*, **8**:3307-3312.
10. Liu, V.F., and Weaver, D.T., 1993. *Mol. Cell. Biol.*, **13**: 7222-7231.
11. Lu, X., and Lane D.P., 1993. *Cell*, **75**: 765-778.
12. Jaspers, N.G.J., Gatti, R.A., Baan, C., Linssen, P.C.M.L., and Bootsma, D., 1988. *Cytogenet. Cell Genet.*, **49**: 259-263.

13. Gatti, R.A., Berkel, I., Boder, E., Braedt, G., Charmley, P., Concannon, P., Ersoy, F., Foroud, T., Jaspers, N.G.J., Lange, K., Lathrop, G.M., Leppert, M., Nakamura,Y., O'Connell, P., Paterson, M., Salser, W., Sanal, O., Silver, J., Sparkes, R.S., Susi, E., Weeks, D.E., Wei, S., White, R., and Yoder, F., 1988. *Nature,* **336**: 577-580.

14. Ziv, Y., Rotman, G., Frydman, M., Dagan, J., Cohen, T., Foroud, T., Gatti, R.A., and Shiloh, Y., 1991. *Genomics,* **9**: 373-375.

15. McConvill, C.M., Byrd, P.J., Ambrose, H.J., Stankovic, T., Ziv, Y., Bar-Shira, A., Vanagaite, L., Rotman, G., Shiloh, Y., Gillett, G.T., Riley, J.H., and Talor A.M.R., 1993. *Hum. Mol. Genet.*, **2**: 969-974.

16. McConvill, C.M., Byrd, P.J., Ambrose, H.J., and Taylor, A.M.R., 1994. *Int. J. Radiat. Biol.*, **66**: S45-S56.

17. Gatti, R.A., Lange, E., Rotman, G., Chen, X., Uhrhammer, N., Liang, T., Chiplunkar, S., Yang, L., Udar, N., Dandekar, S., Sheikhavandi, S., Wang, Z., Yang, H.-M., Polikow, J., Elashoff, M., Teletar, M., Sanal, O., Chessa, L., McConville, C., Taylor, M., Shiloh, Y., Porras, O., Borresen, A.-L., Wegner, R.-D., Curry, C., Gerken, S., Lange, K., and Concannon, P., 1994. *Int. J. Radiat. Biol.*, **66**: S57-S62.

18. Savitsky, K., Bar-Shira, A., Gilad, S., Rotman, G., Ziv, Y., Vanagaite, L., Tagle, D.A., Smith, S., Uziel, T., Sfez, S., Ashkenazi, M., Pecker, I., Frydman, M., Harnik, R., Patanjali, S.R., Simmons, A., Clines, G.A., Sartiel, A., Gatti, R.A., Chessa, L., Sanal, O., Lavin, M.F., Jaspers, N.G.J., Taylor, A.M.R., Arlett, C.F., Miki, T., Weissman, S.M., Lovett, M., Collins, F.S., and Shiloh, Y., 1995a. *Science,* **268**: 1749-1753.

19. Savitsky, K., Sfez, S., Tangle, D.A., Ziv, Y., Sartiel, A., Collins, F.S., Shiloh, Y., and Rotman, G., 1995b. *Hum. Mol. Genet.*, **4**: 2025-2032.

20. Cafferkey, R., Young, P.R., McLaughlin, M.M., Bergsma, D.J., Koltin, Y., Sathe, G.M., Faucette, L., Eng, W-K., Johnson, R.K., and Livi, G.P., 1993. *Mol. Cell. Biol.*, **13**: 6012-6023.

21. Kunz,J., Henriquez, R., Schneider, U., Deuter-Reinhard, M., Movva, N.R., and Hall, M.N., 1993. *Cell,* **73**: 585-596.

22. Hari, K.L., Santerre,A., Sekelsky, J.J., McKim, K.S., Boyd, J.B., and Hawley, R.S., 1995. *Cell,* **82**: 841-847.

24. Hartley, K.O., Gell, D., Smith, G.C.M., Zhang, H., Divecha, N., Connelly, M.A., Admon, A., Lees-Miller, S.P., Anderson, C.W., and Jackson, S.P., 1995. *Cell,* **82**: 849-856.

25. Greenwell, P.W., Kronmal, S.L., Porter, S.E., Gassenhuber, J., Obermaier, B., and Petes, T.D., 1995. *Cell,* **82**: 823-829.

26. Morrow, D.M., Tagle, D.A., Shiloh, Y., Collins, F.S., and Hieter, P., 1995. *Cell,* **82**: 831-840.

27. Imai, T., Seki,N., Saito, T., Yamauchi, M., Matsuda, Y., Ito, H., Ogiwara, A., Nomura, N., and Hori, T., 1995. *DNA Res.*, **2**: 113-121.

28. Imai, T. and Olson, M., 1990. *Genomics,* **8**: 297-303.

29. Imai, T., Iida, A., Miwa, T. et al., 1990. *Methods: A Companion to Methods in Enzymol.*, **1**: 180-185.
30. Imai, T., Yamauchi, M., Seki, N. et al., 1996. *Genome Res.*, **6**: 439-447.
31. Valdes, J.M., Tagle, D.A., and Collins, F.S. 1994. *Proc. Natl. Acad. Sci. USA*, **91**: 5377-5381.
32. Nakai,K., and Kanehisa, M., 1992. *Genomics*, **14**: 897-911.
33. Nigg, E.A., 1993. *Trends Cell Biol.*, 3: 296-301.
34. Kishimoto, A., Nishiyama, K., Nakanishi, H., Uratsuji, Y., Nomura, H., Takeyama, Y., and Nishizuka, Y., 1985. *J. Biol. Chem.*, **260**: 12492-12499.
35. Woodgett, J.R., Gould, K.L., and Hunter, T., 1986. *Eur. J. Biochem.*, **161**: 177-184.
36. Feramisco, J.R., Glass, D.B., and Krebs, E.G., 1980. *J. Biol. Chem.*, **255**: 4240-4245.
37. Cross, S.H., Charlton, J.A., Nan, X., and Bird, A.P., 1994. *Nature Genet.*, **6**: 236-244.
38. Thiesen, H.-J. and Bach, C., 1990. *Nucleic Acids Res.*, **18**: 3203-3209.
39. Fernandes, M., Xiao, H. and Lis, J.T., 1994. *Nucleic Acids Res.*, **22**: 167-173.
40. La Thangue, N.B., 1994. *Trends Biol. Sci.*, **19**: 108-114.

15. Radioadaptation of Mutation in Human x Hamster Hybrid A_L Cells

Akiko Ueno, Diane Vannais, Megan McGraw, Russell Drabek, Daniel Gustafson, Jeanie Robinson and Charles Waldren

Department of Radiological Health Sciences (RHS-MRB), Colorado State University, Fort Collins, CO 80523, USA

Introduction

Cancer is a genetic disease driven by the accumulation of mutations in a cell and its progeny e.g {479, 3452}. (A Mutation is defined as any heritable genetic change). Chromosomal mutations involving specific rearrangements of large pieces of chromosomes or loss or gain of whole chromosomes are at least as important in carcinogenesis as are intragenic or point mutations {479, 3452}. Cells in many cancers have lost the ability to maintain the integrity of their chromosomes so that high levels of chromosomal rearrangements are a hall-mark of tumorigenesis. Although repair genes, transcription factors, and cell cycle regulatory factors such as p53 and Raf-1 have been implicated in the maintenance of genome stability, e.g {3265, 4174, 4171, 4182}, how stability is maintained or instability engendered is not well understood.

Besides being a classic mutagen and clastogen, e.g. {2960}, ionizing radiation can cause heritable genomic instability in cells exposed in vitro or in vivo, e.g. {4181, 1275, 3570, 1100, 1488, 4178, 4171, 4182}. On the other hand, considerable evidence indicates that low doses of ionizing radiation (10 cGy or less) can reduce the quantity of chromosomal aberrations and chromosomal mutations induced by a subsequent, higher dose {3596, 3298, 303, 3303, 4015, 1978, 4005, 4068}. Besides affecting mutant quantity, adaptation can also affect the quality of mutants induced at specific loci {1978, 4005, 4068}.

We now find that a low dose of ^{137}Cs-γ rays (4 cGy) significantly decreases (by about half) the yield in human-hamster hybrid A_L cells of hprt- and $S1^-$ mutants induced by a subsequent dose of 3 or 4 Gy. Further, the spectrum of S1- mutants induced by the low + high dose regimen is much different from for the high dose alone. The most obvious, and perhaps most interesting change in the spectrum is the large number of complex, unstable mutants with the dual-dose regimen compared to the high dose alone.

Methods and Materials

The A_L human-hamster hybrid cell line and its application in mutation analysis have been described {78, 3062, 4166, 4068}. A_L cells contain a standard set of Chinese Hamster Ovary (CHO) chromosomes plus a single copy of human chromosome, number 11. The MIC1 gene at 11p15.5 encodes the S1 human cell surface antigen. Wild type ($S1^+$) cells are killed by serum complement + monoclonal anti S1 antibody. $S1^-$ cells survive to form colonies which can be counted to give the fraction of $S1^-$ mutants a population. The mutant fraction, M_f = number of colonies/number of cells inoculated, corrected for the cloning efficiency in complement alone. M_f are expressed for convenience as mutants per 10^5 clonable cells. Induced mutant fractions, IM_f, are obtained by subtracting background mutant fractions from the total number of mutants.

Cell culture and irradiation conditions were as described {3062, 4166, 4068}. Briefly, exponentially growing cells were exposed at room temperature in a MARK I ^{137}Cs-γ Cs-gamma irradiator (J.L. Shepherd & Associates, Glendale, CA), at a dose rate of 1 Gy/min for higer doses, 20 cGy/min for that adaptive dose. Cell survival for each exposure condition was determined by a standard colony-forming assay. Cohorts containing at least 5 x 10^5 surviving cells were incubated for 10 days for mutant expression, then "challenged" with antisera + complement to determine the fraction of $S1^-$ mutants. Each population was evaluated at least three times. Yields $hprt^-$ mutants, defined by their resistance to 6-thioguanine, were determined as described {78}.

Genotypes of $S1^-$ mutant clones were defined by Southern or polymerase chain reaction (PCR) analysis of DNA extracted from populations of cells expanded from individual S1- clones using a series of DNA probes that map along chromosome 11 to determine marker loss or presence {4068, 4166}.

Results and Discussion

Cell Killing and Mutation

Survival and mutation were determined for cells treated as follows: (a) unirradiated (controls). (b) 4 cGy ^{137}Cs-γ (adaptive dose). (c) 4 Gy, or (d) 4 cGy followed by a "challenge dose" of 3 or 4 Gy. As shown in Table 1, the 4 cGy adaptive dose did not affect survival with the challenge dose or by itself, nor did it alter the growth rate of cells or rate of entry or distribution of cells in the compartments of the cell cycle (latter data not shown).

But, as shown in Table 2, the adaptive dose caused a significant decrease (by about 50%) in the yield of S1- mutants induced by the challenge dose. Adaptation similarily decreased the number of hprt- mutants from $IM_f = 2 \pm 1$ for 3 Gy alone to $1 \pm .04$ for 4 cGy + 3 Gy, but the reduction we observed was not statistically significant because the 100-fold lower yield per unit dose at the hprt locus makes it harder to establish significance. The adaptive dose was not mutagenic of itself.

Table 1. Effects of an adaptive dose on survival of A_L cells

γ-ray Dose (Gy)	Relative Surviving Fraction
0	1.0
0.04	1.09 ± 0.02
4	0.2 ± 0.04
0.04 + 4	0.19 ± 0.02

Data are means ± SD of at least 6 experiments. The absolute plating efficiency for unirradiated cells averaged 80%.

Table 2. Effect of adaptation on the yield of $S1^-$ mutants

Dose (Gy)	Induced M_f
0	0
0.04	3 ± 1
4	214 ± 16
0.04 + 4	111 ± 19

Data are means ± SD for at least 6 experiments.

Effects of cycloheximide (CX), actinomycin-D and 3-aminobenzamide (3- AB)

It has been reported that the adaptive decrease observed in chromosome aberrations was diminished by CX, actinomycin-D or by 3-AB present between the doses of radiation, indicating a role of protein synthesis and mRNA synthesis, and for poly (ADP) ribose polymerase activity in this response {4028,3718}. We have found (Table 3) that these two agents also affected the adaptive response of S1- mutation: CX and 3-AB reduced the radioadaptive effect on mutant yields by 95% and 80%, respectively. At the doses used, neither CX nor 3-AB was mutagenic and they had little effect on cell survival. Actinomycin-D, on the other hand, was itself so lethal and mutagenic (data not shown) that evaluating its effect on adaptation was not possible. Its genotoxicicity is presumably due to its interactions with DNA {4176}.

Effect of adaptation on mutant spectra

In so called "simple" S1- mutants, the markers on chromosome 11 are lost in a continuous fashion extending from the MIC1 locus {4068,4166}. In "complex", on the other hand, the loss of markers appears to be intermittent and discontinuous along the chromosome {4068,4166}. The ratio of simple to complex mutants depends on the quality of a particular radiation and on the regimen of exposure for the same radiation. We found, for example, relatively few complex mutants, about 3%, in the mutant spectra for high dose rate ^{137}Cs-γ rays compared to spectra for a low dose rate (a few cGy/hr), high dose rate + caffeine, high dose rate exposures to HZE Fe and other high LET irradiations, or for spontaneous spectra in all of which at least 25% of the mutants are complex {3062, 4166}.

Table 3. Effect of 3-AB and CX on yields of S1⁻ mutants

Treatment	Relative yield of S1⁻ mutants
4 Gy	100
4 cGy + 4 Gy	46
(" ") + 3-AB	80
(" ") + CX	95

3-AB and CX were used at 5 mM and 10 μg/ml respectively.

We now find that the incidence of complex mutations induced by acute, high dose rate ^{137}Cs-γ rays is increased almost 30 fold by the adaptive dose (Table 4). Adaptation was previously reported to alter the spectrum of hprt- mutants, mainly by decreasing the proportion of deletion mutants {1978,4005}.

The adaptive dose also increased (Table 4) the incidence of complex S1-mutants in populations exposed to only that dose compared with unirradiated, control populations.

Table 4. Relative proportion of simple and complex S1⁻ mutants.

γ-ray Dose, Gy	Number Mutants Analyzed	Percent	
		Simple	Complex
0	94	72	28
0.04	23	22	78
4	75	97	3
4 + 0.04	28	11	89

Although most of the S1- mutants generated by the adaptive and challenge dose regimen were complex (about 89%), complex mutants do not constitute a large proportion of the total cell population. For example, per 10^5 cells at risk, the combined doses, adaptive + challenge, add about 111 mutants (~100 of which are complex) to the 20 (~ 6 complex) mutants preexisting in this population. The fraction of complex mutants per 105 cells is, then, [0.89(111) + 0.28(20)] ~100/105 = 0.1%. This incidence of complex S1- mutants is much lower than in those cases where instability occurred in as many as 30% of the surviving cells {3570,1488} and may, therefore, result from different mechanisms.

"Complex mutants" are cytogenetically complicated and unstable

Preliminary cytogenetic analysis by classical banding and by fluorescence in situ hybridization (FISH) techniques {2960} indicates that populations of $S1^-$ mutants defined as complex by molecular analysis are, in fact, not comprised of a single population of cells each of which has lost the same set of intermittent markers. Rather, these populations are nonclonal and contain several (10 or less) subpopulations each with a different, but specific, set of chromosomal rearrangements including deletions in chromosome 11 and translocations or insertions of fragments of chromosome 11 onto or into CHO chromosomes. Furthermore, periodic cytogenetic analysis of populations derived from complex mutants revealed that the kinds of rearrangements and the proportion of cells with a given rearrangement changed during culture periods of a few weeks. Thus these complex mutants displayed a hallmark of transmissible genomic instability (TGI). This situation is in contrast to populations grown up from simple mutants which are largely made up a single population of cells of a single karyotype which remains stable for many weeks in culture.

We have reported that caffeine acts to increase the fraction of complex mutants in irradiated populations {3062}, which implicates perturbations of cell cycle progression, interference with repair, and possible missegregation of immature chromosomes as factors in instability. We did not, however, observe any effect of the adaptive dose on the cell cycle so that the increased instability appears not due to cell cycle perturbation. Our working hypothesis is that the adaptive dose affects recombination repair processes, possibly via effects initially mediated by the cell membrane, in ways that increase translocation and other chromosomal rearrangements created by subsequent damages, spontaneous or induced. Since this destabilization is heritable, it follows that its regulation must be under genetic control so that isolating regulatory mutants should be possible. In this regard, we have recently isolated from A_L a mutant cell line (A_L-179) {4166} in which about 25% of the $S1^-$ mutants induced by a single high dose rate exposure to ^{137}Cs-γ are complex compared to 3% for A_L cells. Genetic characterization of A_L-179 should shed light on processes that normally act to maintain genomic stability.

Acknowledgments

Supported by National Institutes of Health Grants CA36447, CA56392, NRSA-CA09236 & CA64039, a joint NASA NSCORT to Colorado State University and the Lawrence Berkeley Laboratory, NASA contract NAG 9501-0232, & the CAW/DBV Lab Discretionary Fund.

Bibliography

1. Rowley JD. Cancer Res 1990; 50:3816-3825.
2. Cavenee WK, White RL. Scientific American 1995; 272:72-79.
3. Greenblatt MS, Bennett WP, Hollstein M, Harris CC. Cancer Res 1994; 54:4855-4878.
4. Kasid U, Suy S, Dent P, Ray S, Whiteside TL, Sturgill TW. Nature 1996; 382:813-816.
5. Murnane J. Mutat Res 1996; 367:115-121.
6. Morgan WF, Day JP, Kaplan MI, McGhee EM, Limoli CL. Radiat Res 1996; 146:247-258.
7. Cornforth MN, Bedford JS. Advan Radiat Biol 1993; 17:423-496.
8. Weissenborn U, Streffer C. Int J Radiat Biol 1988; 54:659-674.
9. Little JB, Gorgojo L, Vetrovs H. Int J Radiation Oncology Biol Phys 1990; 19:1425-1429.
10. Kadim MA, Macdonald DA, Goodhead DT, Lorimore SA, Marsden SJ, Wright EG. Nature 1991; 355:738-740.
11. Kadhim MA, Macdonald DT, Goodhead DT, Lorimore SA, Marsden SJ, Wright EG. Nature 1992; 355:738-740.
12. Sabatier L, Durtillaux B, Martin MB. Nature 1992; 357:548
13. Kadim MA, Lorimore SA, Townsend KM, Goodhead DT, Buckle VJ, Wright EG. Int J Radiat Biol 1995; 67:287-293.
14. Olivieri G, Bodycote Y, Wolff S. Science 1984; 223:594-597.
15. Shadley JD, Afzal V, Wolff S. Radiat Res 1987; 111:511-517.
16. Wolff S, Afzal V, Wiencke JK, Olivieri G, Michaeli A. Int J Radiat Biol 1988; 53:39-48.
17. Kelsey KT, Memisoglu A, Frenkel D, Liber HL. Mutat Res 1990; 263:197-201.
18. Joiner MC. Int J Radiat Biol 1994; 65:79-84.
19. Rigaud O, Papadopoulo D, Moustacchi E. Radiat Res 1993; 133:94-101.
20. Rigaud O, Laquerbe A, Moustacchi E. Radiat Res 1995; 144:181-189.
21. Ueno AM, Vannais DB, Gustafson DL, Wong JC, Waldren CA. Mutat Res 1996 (In press).
22. Waldren C, Correll L, Sognier MA, Puck TT. Proc Natl Acad Sci USA 1986; 83:4839-4843.
23. McGuiness S, Shibuya M, Ueno A, Vannais D, Waldren C. Radiat Res 1995; 142:247-255.
24. Waldren C, Vannais D, Drabek R, Gustafson D, Kronenberg A, Ueno A. Adv Space Res 1996 (In press).
25. Kumar A, Kiefer J, Schneider E, Crompton NEA. Int J Radiat Biol 1985; 47:103-112.
26. Wolff S, Wiencke JK, Afzal V, Youngblom J, Cortes F. In: Baverstock KF, Stather JW, eds. Low Dose Radiation: Biological bases of risk assessment. London: Taylor and Francis, 1989:446-454.
27. Kersten W. Biochim Biophys Acta 1961; 611

16. Mutation as a (Bio)Chemical Process (1958-1996)

Bernard S. STRAUSS

Department of Molecular Genetics and Cell Biology,
The University of Chicago, Illinois 60637, USA

I. Mutation Studies before 1960

In 1958 we were sure that DNA was the genetic material (1). The discovery of DNA transformation (2), was followed by the discovery of a molecular structure (3) which permitted investigators to understand how a compound as monotonous as DNA, with only four repeated elements, could encode a large number of complex messages. Given this understanding, the experiment of Hershey and Chase (4) was seen as convincing, notwithstanding any of the arguments about contaminating protein which had previously dogged the studies on DNA transformation (5). Although we did not know the details, a triplet code was already considered most likely. That genetic information resided in the sequence of DNA bases was evident, but there was only a small library of sequenced protein and none of DNA so that elimination of an overlapping code was just barely possible (6). We did know that protein synthesis occurred in the cytoplasm and that RNA was involved, but the distinction between the different forms of RNA had not been made and it took a decade to discover that cells did not contain ribosomes specific for each protein (7).

It was possible to show that not all portions of the genome were equally mutable, i.e. there were hot spots (8). Freese (9) used the base analog bromouracil to demonstrate the differential revertibility of mutations arising from different initial treatments and deduced that point mutations were of two types which he named transitions and transversions. The changes he thought of as transversions were recognized to be frameshifts a few years later (10). We knew that chemicals as well as radiation could induce mutation (11, 12) and that different compounds produced relatively different mutagenic effects in different genes (13). It was already recognized that "target theory" in its most extreme form could not be correct. This theory supposed that the lethal and mutagenic effects of radiation could be understood in purely physical terms as the result of radiation "bullets" hitting, and inactivating, a biological target (14). The theory had numerous successes, particularly in calculating the size of dried virus from the radiation dose required to inactivate. However, by 1958 there was already good evidence that it was an oversimplification. Hershey, and later Stent had shown that ^{32}P decay was lethal when the isotope was incorporated into DNA, but the lethality was greatly decreased when the decay occurred in growing cells (15). The effects of non-ionizing radiation were clearly mediated by metabolic systems. Doudney and Haas (16) had described a process they called mutation frequency decline in which the yield of mutants depended on just how long cells were allowed to incubate in minimal medium after treatment and Witkin had pointed out the importance of protein synthesis in mutation (17). The experiments of Kelner on photoreactivation showed that radiation damage was reversible for some period after exposure (18). For all these reasons it was possible to suppose that mutation was a chemical or a biochemical process, involving metabolic systems which responded to damaged DNA (1).

II. A Decade of Discovery

The 1960's produced an array of important discoveries which set the pattern for our current understanding of biology (see Stent (7) for a historical account) . Within a short time the details of the universal genetic code were unraveled, including the discovery that the code was highly degenerate. About 25 percent of random changes in the coding region of a gene produce no amino acid change, and are so-called "silent mutations". The roles of the different classes of RNA in protein synthesis were unraveled along with the first mechanistic interpretations of the mode of gene regulation. A "central dogma" of molecular biology which presumed that the flow of information went from DNA to RNA to protein was revised with the discovery of reverse transcriptase. The 1960's were a euphoric period for molecular biologists with the principles becoming clear and the details yet to come.

Mutation studies were revolutionized by the discovery of the repair pathways for damaged DNA and by the realization that not only damaged but mismatched DNA's were repaired (19). Critical to this understanding was the recognition of mutant organisms with enhanced sensitivity to radiation and chemicals, first in bacteria (20) and later in humans(21). The sensitivity of such organisms to radiation showed that metabolic processes were critical in the mediation of radiation and chemical damage. Detailed knowledge of the processes in prokaryotes and eukaryotes came with the exploitation of the cloning and sequencing methodologies of the 1970's. In the late 1960's and early 1970's there were often hints that cells could recognize that their DNA was damaged and withdraw from the cell cycle until the damage was repaired but such interpretations of the data seemed highly teleological. It is only recent studies on the cell cycle and on the phenomenon of programmed cell death (apoptosis) which have made it clear that such mechanisms not only exist but play a critical role in the response of cells to damage. Understanding how mutations originate requires not only an understanding of alterations in DNA structure and the response of polymerase to the alterations but also a consideration of the interactions of transcriptional state, cell cycle control, apoptotic death, and DNA repair mechanisms (Fig. 1).

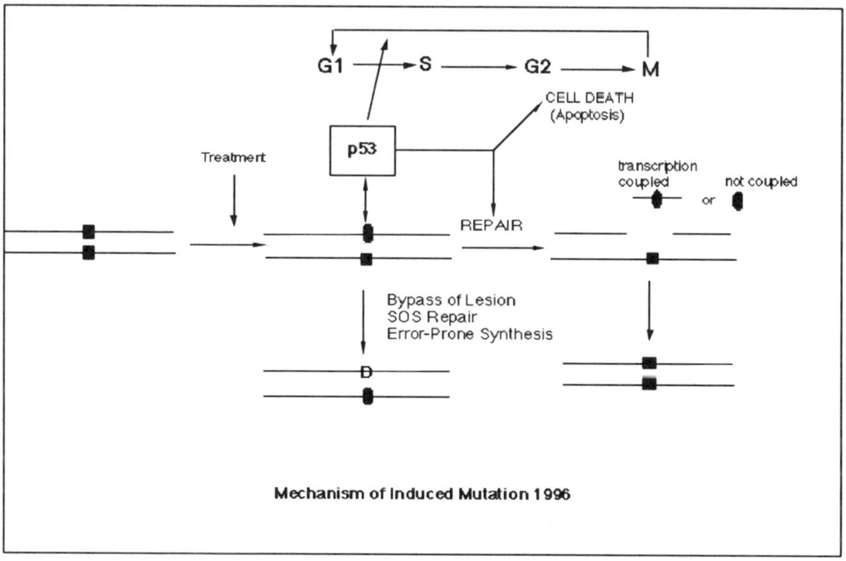

Fig 1. A current view of factors determining the appearance of mutations.

III. Current Problems in Mutation Research

Current studies on mutation fall into two major categories. First we need to know the details of many mechanisms we can only guess at in outline. Among such problems are included the details of how DNA polymerases and their accessory proteins manage to synthesize past lesions in the DNA. An understanding of the mechanistic details of the interactions of DNA structure, type of lesion and polymerase will eventually come from systematic analysis (22) as will recognition of the important endogenous mutagens and oxidation products.

Hypermutability in Immunoglobin Production:

Analysis of three, possibly related, phenomena may reveal new insights. The first of these is the hypermutability which accompanies the generation of diversity in antibody formation. Following V(D)J joining and in response to antigen stimulation, particular cells in humans (and in mice) respond by mutating. Cells with mutated immunoglobins either have higher affinity for the stimulating antigen, in which case they are selected, or lower affinity, in which case their apoptotic mechanisms are activated and the cells die. The mutations observed are mostly point mutations with rare deletions or insertions of single nucleotides. Bacterial sequences, a globin gene or artificial substrates can be somatically mutated when inserted into an appropriate context indicating that the sequence which is actually mutated does not itself initiate the process of mutation. Hotspots of mutation are frequently observed at different positions in different sequences. Mutations upstream of the transcriptional start site are extremely rare and there is no defined 3' end point of the mutation tract but the observed mutations extend only over about 1.5 kb (23). This hypermutability is limited to certain stages in development. About 15 percent of the mutations in the complementarity determining region are silent suggesting that the mutational change is random and that the mutated cells are then subject to selection (24). Hypermutability is not the only way to generate diversity, even in related species. While mice and humans mutate, chickens and rabbits copy sequences already present in the genome to generate diversity by a process of gene conversion. It has been suggested that the association of the hypermutability process with transcription implicates transcription coupled repair. Mutation is hypothesized to occur as a consequence of the DNA synthesis accompanying repair (23). This hypothesis requires that there be a factor, present only in B cells, which loads onto a transcription complex. Although there is little current interest in the hypothesis that reverse transcriptase is involved (25), this enzyme is involved in the movement of retrotransposons and recent studies implicate the reverse transcriptase in the repair of double strand breaks, key intermediates in the generation of antibodies (26, 27). Since retrotransposition is a process with a high error rate (28), in part due to the lack of proofreading capacity of the reverse transcriptase (29, 30, 31), it seems reasonable to reexamine the possible role of reverse transcriptase in immunoglobin hypermutability.

Genetic Instability and Carcinogenesis:

Changes in chromosome number and structure are numerous in carcinogenesis. Gene amplification occurs and indeed, appears to be restricted to cells undergoing transformation (32). Numerous point mutations are found. A proposed mechanism of carcinogenesis for which there is substantial evidence requires the accumulation of multiple genetic changes (33). However, a closer look indicates that there is relatively little evidence for gene hypermutability and even less understanding of what the mechanism might be. The question is whether tumors are hypermutable for the generation of point mutations. That point mutations do accumulate has been demonstrated and specific mutations can play an etiologic role in tumor development (34). However, such demonstrations do not establish the fact of hypermutability since mutations produce clones with a selective advantage and the growth of such clones reflects selection as much as it does mutation. Given the normal germinal spontaneous mutation rate it would not be possible to accumulate the number of mutations found or required in tumors in the number of cell generations available (35). However, the numerical argument requires that mutations accumulate in one cell before a malignant transformation occurs. An alternative model which posits the sequential accumulation of mutations in a series of clones, one derived from the other, permits the almost unlimited accumulation of mutations within a lifetime. However, such models require that each mutation provide a sufficient selective advantage so that a clone containing that mutation does grow to large enough size to undergo another "favorable" mutation.

Some cancer prone families have mutations in mismatch repair genes which increase the overall mutation rate (36). However, not all tumor derived lines have a higher mutation rate than their normal counterparts (37) which implies that hypermutability need not be a permanent attribute of tumors. Secondly, individuals completely deficient in mismatch repair have been observed and although they do have tumors, they do not have the massive load of tumors that might be expected (38). In addition, not all kinds of tumors occur at high frequency in colon cancer prone families. Mismatch repair is only one of several repair processes which protect against point mutations. The proofreading exonuclease, which accompanies DNA polymerase, is at least as efficient (39). There is some specialization so that mismatch repair is most efficient in the surveillance of errors at multiple nucleotide repeats, whereas proofreading is more efficient at detecting isolated point mutations (Table 1). It may be that particular genes with repeat elements are the targets in individuals deficient in mismatch repair (40). Whether there is a general hypermutability required for tumor generation remains an open question.

Table 1. Role of Different Surveillance Systems in the Correction of Errors

Type of Mutation	Percent Errors Corrected by: Mismatch Repair	Proofreading
Base Substitutions	<1	>99
Frameshift $(CA)_2$	1	99
Frameshift (8 C)	20	80
Frameshift $(CA)_{14}$	60	40

Data on Base substitutions from Schaaper (39). Data on frameshifts from Strauss, Sagher and Acharya, *unpublished*.

The regulatory protein p53 has been much studied and over 5000 mutations in tumors have been tabulated (41, 42). The mutations are distributed throughout the p53 gene with concentrations at sites that are important for interactions with other proteins and/or DNA. The p53 protein forms tetramers and many mutations are dominant because even one mutant subunit can inactivate normal function (43). Subsequent mutations may have no additional effect. The data base records mutations from 4,673 tumors after subtracting the 417 cell lines included. Among these are 149 silent mutations. Does the occurrence of silent mutations tell us anything about the process of mutagenesis in carcinogenesis?

If mutation were random with no selection, about 25 percent of all mutations should be silent. If only the first p53 mutation results in a selective advantage and subsequent mutations have (almost) no additional effect, an examination of multiple p53 mutations in tumors should give a better picture of the mutation process before selection (Table 2). About twenty percent of tumors with multiple mutations include at least one which is silent. The percentage of silent mutations among the multiples (after correction for the first mutation required to enter the sample into the bin) is 18 percent. This frequency of silent mutations is similar to the percent of silent mutations found in immunoglobin (14.6%) (24) and is close to what what would be expected if random mutagenesis had occurred. Of course, "silent mutations" need not be selectively neutral. There does seem to be selection for particular codons in *Drosophila*, in yeast and in *E. coli*, but not in humans. Similar human proteins

Table2.

p53 Numerology (42)		
1. Total tumors with mutations	4,673	
Silent mutations in tumors (Percent)	149 (3.2)	
2. Total tumors with multiple mutations recorded	247	
Percent of tumors with multiple mutations at least one of which is silent = 49/247) x 100		19.8
3		
total silent mutations (tumors)	58	
total mutations (tumors)	566	
percent silents corrected for ascertainment bias (58/566-247) x 100	18.2	
4. Corrected percentage of all mutation in the 4673 tumors with mutations = (566 - 247/ 4673) x 100 =	6.8	

made in large amount at the same time may have widely different codon usage, indicating non-selectivity (44, 45). The silent mutations listed in the tabulation do not occur at the sites of previously reported polymorphisms (42) and can be assumed to have occurred during tumorigenesis. The measured mutation frequency for one mutable p53 codon (number 249) has been estimated at between 2 - 13 x 10^{-7} per base in non-tumor tissue (46). This estimate seems very high but even if it were assumed that all nucleotides had the same mutability this would give a <u>maximum</u> frequency of only 1179 x 10^{-7} = 1.5 x 10^{-3} or 0.15%. The silent mutation frequency in p53 from tumors is therefore <u>at least</u> 25 times greater than expected.

If there are multiple independent mutational events, the expected frequency of multiple mutations can be calculated using the Poisson distribution. Given the number of tumors with two or three p53 mutations (195 and 43) it is possible to calculate the expected number with four and five (1 and 0) but there are far too many in the data set with only one mutation (4,673 compared to 582 expected). There are several possible reasons for this discrepency. An important, perhaps critical, reason is PCR error. A second possibility is that the number of tumors with multiple mutations is underestimated because the data is based on a set of determinations in which only a subset of the possible mutations are observed. Introns are not sequenced and only 99 of the 550 publications on which the data are based report sequencing the entire coding sequence, that is codons 2 to 11. Yet another possibility has been suggested by John Cairns who made the Poisson calculation (personal communication) and who supposes as one interpretation that the data fit the hypothesis of a subset of about 10 percent hypermutable tumors. The location of the mutations suggests yet another explanation, namely that the mutations do not occur independently but rather as the result of multiple events (46). Of 247 multiple mutations (mostly doubles) , 76 or 31 percent are within ten codons of each other. Considering that in most cases only exons 5-8 (181 amino acids) or 4-9 (299 amino acids) have been sequenced this is not an overwhelming excess since one would expect such a juxtaposition by chance about ten percent of the time. However, out of the 43 triple mutations, at least six have mutations clustered close to one another in a manner suggestive of multiple errors in a single mutational event, e.g. tumor TU-33 a bladder tumor with changes in codons 135, 138 and 140 and bases 403, 412 and 418.

The published data therefore suggest that there is a high rate of non-selected mutation in the p53 gene during tumor development. The mutagenic process may be special in that multiple mutations occur close together. The existence of silent mutations implies that not all the mutations observed in tumors need have an etiologic role. What is not clear is whether the high mutability of p53 is indicative of the unique nature of p53 and perhaps a few other genes. High mutability is not found in all genes examined in tumors. An analysis of 387 tumor samples for WAF-1 mutations by SSCP analysis failed to disclose any new mutations

(47). It may be that the structural properties of p53 permit mutations to accumulate and be detected. On the other hand, it may be that p53 is uniquely mutable in somewhat the same way that the immunoglobulin gene is mutable.

Adaptive Mutation:

The "Lamarckian Heresy", the hypothesis that mutations are not only selected but occur as a result of environmental direction continually recurs in biology, possibly as a result of our instinctive rejection of a Universe governed by chance. The most recent reincarnation of the hypothesis is due to Cairns (48) who very properly pointed out that the evidence for random preexisting mutations in populations comes from the use of lethal selection in the analysis. Cairns suggested that non-lethal selection would permit the production of mutations leading to adaptive growth. *Escherichia coli* with a revertible inactivating mutation in the ß-galactosidase gene were permitted to incubate on agar plates with lactose present. Lactose-utilizing mutants appeared. Other mutations that did not provide a selective advantage were not detected in the early experiments. Soon after Cairns' initial report it was recalled that similar experiments had been done by Francis Ryan (49, 50, 51) who showed that histidine-requiring bacteria incubated in a non-growing state could mutate to histidine-independence with no evidence for major DNA synthesis or cell turnover accompanying the mutation. These results have provoked much controversy and much interesting experimentation. The most likely artifacts have been eliminated and some special aspects of the experimental systems used have been recognized . For example, one of the favorite systems requires a frameshift mutation to occur in a ß-galactosidase gene located on an episome. Both the recombination system and the system(s) involved in the conjugational transfer of episomes have been shown to be involved in the reversion (52, 53, 54, 55, 56). Current explanations invoke localized DNA synthesis combined with a failure of mismatch repair in the non-dividing population. Two laboratories have recently reexamined the critical question as to whether mutations occur in non-selected genes when organisms are kept under non-lethal selection. Both have now found that mutations do occur in non-selected as well as selected genes (57, 58). Both laboratories suggest that a hypermutable state has been induced although one group supposes this may be confined to the episome (57). The second group finds hypermutability throughout the genome but supposes it limited to a small proportion of the cells under the stress of selection (58).

It appears that the phenomenon described by Ryan and later by Cairns is now accessible and amenable to genetic investigation although it needs to be pointed out that the recent studies relate to only one experimental system. Current data indicate that organisms under stress become hypermutable. It is a hypothesis that the three phenomena discussed above, immunoglobin production, carcinogenesis and "adaptive mutation" in bacteria share a property of hypermutability for a restricted period in the life cycle. It will be interesting to see whether the three different phenomena have any mechanistic similarities.

Acknowledgment. This paper is dedicated to the memory of Professors Francis Ryan and Hideo Kikkawa both of whom played major roles in developing and fostering my interest in mutation and in Japan. The speculations on p53 mutability were taken from a more comprehensive manuscript (in preparation). I would like to thank Dr. Monica Hollstein and Dr. John Cairns for their generous help and critical comments on this subject. However, this discussion reflects only my own views. The experimental work reported from my laboratory is taken from a recently submitted manuscript and was supported by a grant from the National Cancer Institute, National Institutes of Health (5R37 CA32436-15).

REFERENCES

(1) Strauss, B. S. *Iden (Heredity)* **1960**, *14,* 28-37.

(2) Avery, O. T.; MacLeod, C. M.; McCarty, M. *J. Exptl. Med.* **1944**, *79,* 137-158.

(3) Watson, J. D.; Crick, F. *Nature* **1953**, *171,* 737-738.

(4) Hershey, A. D.; Chase, M. *J. Gen. Physiol.* **1952**, *36,* 39-56.

(5) Mirsky, A. E. in *Some chemical aspects of the cell nucleus*; Dunn, L. C.; MacMillan, New York, 1951; pp 127-153.

(6) Gamow, G.; Rich, A.; Ycas, M. *Advan. Biol. Med. Phys.* **1956**, *4,* 23-68.

(7) Stent, G. S. *Molecular genetics. An introductory narrative.*; W.H. Freeman and Co.: San Francisco, 1971; pp xvi +650 .

(8) Benzer, S.; Freese, E. *Proc. Natl. Acad. Sci. USA* **1958**, *44,* 112-119.

(9) Freese, E. *Proc. Natl. Acad. Sci. USA* **1959**, *45,* 622-633.

(10) Brenner, S.; Barnett, L.; Cfrick, F. H. C.; Orgel, A. *J. Mol. Biol.* **1961**, *3,* 121-124.

(11) Auerbach, C.; Robson, J. M. *Nature* **1946**, *157,* 302.

(12) Auerbach, C. *Cold Spring Harbor Symp. Quant. Biol.* **1951**, *16,* 199-213.

(13) Westergaard, M. *Experientia* **1957**, *13,* 224-234.

(14) Lea, D. E. *Actions of Radiations on Living Cells*; Cambridge Univ. Press: Cambridge, 1946; pp .

(15) Stent, G. *J. Gen. Physiol.* **1955**, *38,* 853-865.

(16) Doudney, C. O.; Haas, F. L. *Proc. Natl. Acad. Sci. USA* **1959**, *45,* 709-722.

(17) Witkin, E. M. *Cold Spring Harbor Symp. Quant. Biol.* **1956**, *16,* 357-372.

(18) Kelner, A. *Proc. Natl. Acad. Sci. USA* **1949**, *35,* 73-79.

(19) Wildenberg, J.; Meselson, M. *Proc. Natl. Acad. Sci. USA* **1975**, *72,* 2202-6.

(20) Hill, R. *Biochim. Biophys. Acta* **1958**, *30,* 636-637.

(21) Cleaver, J. E. *Nature* **1968**, *218,* 652-656.

(22) Bebenek, K.; Beard, W. A.; Casas-Finet, J.; Kim, H.; Darden, T. A.; Wilson, S. H.; Kunkel, T. A. *J. Biol. Chem.* **1995**, *270,* 19516-23.

(23) Peters, a.; Storb, U. *Immunity* **1996**, *4,* 57-65.

(24) Kaartinen, M.; Kulp, S.; Makela, O. in *Characteristics of selection-free mutations and effects of subsequent selection*; Steele, E. J.; CRC Press, Boca Raton, 1991; pp 105-114.

(25) Steele, E. J.; Pollard, J. W. *Molecular Immunology* **1987**, *24,* 667-73.

(26) Moore, J. K.; Haber, J. E. *Nature* **1996**, *383,* 644-646.

(27) Teng, S.-C.; Kim, B.; Gabriel, A. *Nature* **1996**, *383,* 641-644.

(28) Boeke, J.; Garfinkel, D. J.; Styles, C.; Fink, G. R. *Cell* **1985**, *40,* 491-500.

(29) Bebenek, K.; Abbotts, J.; Wilson, S. H.; Kunkel, T. A. *J. Biol. Chem.* **1993**, *268,* 10324-34.

(30) Shen, J. C.; Creighton, S.; Jones, P. A.; Goodman, M. F. *Nucleic Acids Research* **1992**, *20,* 5119-5125.

(31) Mansky, L. M.; H.M., T. *J. Virol.* **1995**, *69,* 5087-5094.

(32) Tlsty, T. D.; Margolin, B. H.; Lum, K. *Proc. Natl Acad. Sci. USA* **1989**, *86,* 9441-9445.

(33) Fearon, E. R.; Vogelstein, B. *Cell* **1990**, *61,* 759-67.

(34) Levine, A. J. *Ann. Rev. Biochem.* **1993**, *62,* 623-651.

(35) Loeb, L. A. *Cancer Research.* **1991**, *51,* 3075-9.

(36) Fishel, R.; Lescoe, M.; Rao, M.; Copeland, N.; Jenkins, N.; Garber, J.; Kane, M.; Kolodner, R. *Cell* **1993**, *75,* 1027-1038.

(37) Kaden, D.; Gadi, I. K.; Bardwell, L.; Gelman, R.; Sager, R. *Cancer Research* **1989**, *49,* 3374-3379.

(38) Parsons, R.; Li, G. M.; Longley, M.; Modrich, P.; Liu, B.; Berk, T.; Hamilton, S. R.; Kinzler, K. W.; Vogelstein, B. *Science* **1995**, *268,* 738-40.

(39) Schaaper, R. M. *J. Biol. Chem.* **1993**, *268,* 23762-23765.

(40) Markowitz, S.; Wang, J.; Myeroff, L.; Parsons, R.; Sun, L.; Lutterbaugh, J.; Fan, R. S.; Zborowska, E.; Kinzler, K. W.; Vogelstein, B.; Willson, J. K. V. *Science* **1995**, *268,* 1336-1338.

(41) Hollstein, M.; Rice, K.; Greenblatt, M. S.; Soussi, T.; Fuchs, R.; Sorlie, T.; Hovig, E.; Smith-Sorensene, B.; Montesano, R.; Harris, C. C. *Nucleic Acids Rsearch* **1994**, *22,* 3551-3555.

(42) Hollstein, M.; Shomer, B.; Greenblatt, M.; Soussi, T.; Hovig, E.; Montesano, R.; Harris, C. C. *Nucleic Acids Research* **1996**, *24,* 141-146.

(43) Wang, P.; Reed, M.; Wang, Y.; Mayr, G.; Stenger, J. E.; Anderson, M. E.; Schwedes, J. F.; Tegtmeyer, P. *Molec. Cell Biol.* **1994**, *14,* 5182-5191.

(44) Sharp, P. M.; Averof, M.; Lloyd, A. T.; G, M.; Peden, J. F. *Phil. Trans. Roy. Soc. Lond. B* **1995**, *349,* 241-247.

(45) Sharp, P. M.; Stenico, M.; Peden, J. F.; Lloyd, A. T. *Biochemical Society Transactions* **1993**, *21,* 835-841.

(46) Harwood, J.; Tachibana, A.; Meuth, M. *Molec. Cell. Biol.* **1991**, *11,* 3163-3170.

(47) Shihara, M.; el-Deiry, W. S.; Wada, M.; Nakamaki, T.; Takeuchi, S.; Yang, R.; Chen, D. L.; Vogelstein, B.; Koeffler, H. P. *Blood* **1994**, *84,* 3781-3784.

(48) Cairns, J.; Overbaugh, J.; Miller, S. *Nature* **1988**, *335,* 142-5.

(49) Ryan, F. J. *J. Gen. Microbiol.* **1959**, *21,* 530-549.

(50) Ryan, F. J. *Proc. Natl. Acad. Sci. USA* **1954**, *40,* 178-186.

(51) Ryan, F. J.; Nakada, D.; Schneider, M. J. *Z. Vererbungslehre* **1961**, *92,* 38-41.

(52) Harris, R. S.; Longerich, S.; Rosenberg, S. *Science* **1994**, *264,* 258-260.

(53) Galitski, T.; Roth, J. R. *Science* **1995**, *268,* 421-423.

(54) Foster, P.; Trimarchi, J. M. *Proc. Natl. Acad. Sci. USA* **1995**, *92,* 5487-5490.

(55) Foster, P. L.; Trimarchi, J. M.; Maurer, R. A. *Genetics* **1996**, *142,* 25-37.

(56) Harris, R. S.; Ross, K. J.; Rosenberg, S. M. *Genetics* **1996**, *142,* 681-691.

(57) Foster, P. **1996**,

(58) Torkelson, J.; Harris, R. S.; Lombardo, M.-J.; Nagendran, J.; Thulin, C.; Rosenberg, S. M. **1996**,

V

Repair of UV-induced DNA Damages

17. Functional Studies of the *XPA* Gene for Nucleotide Excision Repair and Ultraviolet Radiation-induced Skin Carcinogenesis

Kiyoji TANAKA

Institute for Molecular and Cellular Biology, Osaka University
Osaka 565, Japan

Nucleotide excision repair (NER) is the major pathway by which a broad spectrum of DNA damage, including pyrimidine dimers induced by UV irradiation, are removed from the genome. The importance of NER has been indicated by studies on an inherited human disease called xeroderma pigmentosum (XP), which is characterized by the hypersensitivity of skin to sunlight, a high incidence of sunlight-induced skin cancer, and neurological complications . Cells from XP patients are hypersensitive to killing by UV irradiation because of a defect in NER. There are seven different genetic complementation groups (A to G) and a variant form in XP. In addition, there are two genetic complementation groups in Cockayne syndrome (CS-A and B), which is a NER-defective hereditary disorder with clinical symptoms of neurodevelopmental delay and dwarfism, but with no increased incidence of skin cancer, and 11 complementation groups (groups 1 to 11) in UV-sensitive, NER-defective, rodent mutant cell lines . In XP and some of these mutant cells, there are defects in the early stages of NER, including the recognition of damaged DNA, generation of dual incisions on either side of the damaged region, and excision of the oligonucleotides containing the damaged site . In CS cells NER of total genomic DNA is normal but the increased rate of repair of the transcribed strand of active genes (transcription-coupled repair) usually seen in repair-proficient cells is absent . So far, *XPA*, *XPB*(*ERCC3*), *XPC*, *XPD*(*ERCC2*), *XPF*(*ERCC4*, *ERCC11*), *XPG*(*ERCC5*), *CSA*(*ERCC8*), *CSB*(*ERCC6*), and *ERCC1* genes have been cloned. The *ERCC* genes are the human genes that can correct the repair deficiency in UV-sensitive rodent mutant cell lines.

In the early stages of NER, many proteins including the XP proteins are known to be involved. The XPA protein (XPA) preferentially binds to UV- or chemical carcinogen-damaged DNA, suggesting that XPA is involved in the damage recognition step of NER. XPB and XPD have helicase activity and are components of the general transcription factor TFIIH . TFIIH was shown to participate in NER as well as in transcription. Possibly, the bi-directional helicase activities of TFIIH serve to locally unwind DNA around the damaged sites. XPG specifically nicks the DNA duplex at the 3' border of a single-stranded DNA region of a bubble structure and cuts on the 3' side of the lesion. ERCC1 and XPF form a complex which has nuclease activity and cuts on the 5' side of the lesion. Using cell-free NER systems, RPA (replication protein A, also known as RFA and HSSB), a multisubunit single-stranded DNA binding protein, was shown to be involved in NER as well as in DNA replication and recombination. XPC was purified as a tight complex with HHR23B (a human homologue of *Saccharomyces cerevisiae* Rad23) which is required in the early stages of NER, although the function of the complex is not clear yet. It was shown that a factor, designated IF7, is also required for the *in vitro* reconstitution of the incision reaction. Obviously protein-protein interactions might play important roles in NER as well as in DNA replication, recombination, and transcription. Actually, the interactions between proteins related to NER have been detected in the yeast two-hybrid system and in *in vitro* binding experiments; XPA and ERCC1, XPA and RPA, XPA and TFIIH, and XPG and RPA. These interactions may function to coordinate the early stages of NER.

We have cloned the XPA gene and analyzed its role in NER processes and UV ray-induced skin carcinogenesis.[1-6, 10, 11)]

Damaged DNA binding activity of the XPA protein [6, 10, 11)]

XPA protein is a zinc metalloprotein consisting of 273 amino acids which binds preferentially to UV- or chemical carcinogen-damaged DNA, suggesting that it is involved in the recognition of several types of DNA damage during nucleotide excision repair processes. We then identified a DNA binding domain of the XPA protein. The region of the XPA protein responsible for preferential binding to DNA damaged by UV or *cis*-diamminedichloroplatinum (II) (cisplatin) is contained within a truncated derivative of the XPA protein, MF122, consisting of 122 amino acids (^{98}Met to ^{219}Phe) and containing

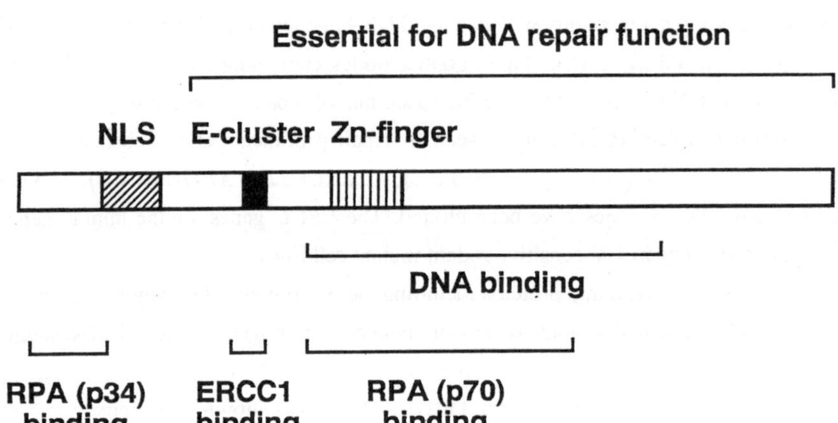

Figure 1. Functional regions of XPA. NLS, nuclear localization signal;cluster, the glutamic acid cluster; Zn-finger, C4 type zinc finger motif.

a C4 type zinc finger motif (Fig. 1). CD (circular dichroism) measurements of the MF122 protein showed that it has a helix rich secondary structure, suggesting that it is a discretely folded, functional mini-domain. ^{113}Cd-NMR analysis of the MF122 polypeptide revealed that ^{113}Cd provided a single signal at a chemical shift value of 709 ppm. This indicated that the ^{113}Cd was tetrahetrally coordinated by four cysteine residues, confirming that the MF122 has a c4 type zync finger motif, and also contains only one zinc atom per protein molecule.

Interactions of the XPA protein with other DNA repair proteins[7)]

The E-cluster and the C-terminal regions are not necessary for the DNA binding activity of XPA. In addition, we have cloned homologues of the human *XPA* cDNA from mouse, chicken, *Xenopus laevis* and *Drosophila melanogaster*. A comparison of the amino acid sequences of these homologues revealed that the N-terminal region is not well conserved except in its nuclear localization signal and E-cluster region, whereas the C-terminal region is highly conserved. These results suggest that the E-cluster and the C-terminal regions may have an important DNA repair function separate from DNA binding. We speculated that they might be domains for protein-protein interactions which coordinate the NER processes. We therefore searched for proteins which bind to XPA using the yeast two hybrid system, and found that the p34 subunit of RPA , ERCC1 and several new proteins bound to XPA.

RPA - XPA interaction [7, 13)]

RPA is a heterotrimer consisting of 70, 34, and 11 kDa subunits and we found that XPA binds the 34 kDa subunit (p34) of RPA using yeast two hybrid system. The direct association between XPA and the RPA complex was confirmed by *in vitro* experiments. We also examined whether the other subunits of RPA bind XPA. The GST-XPA beads were mixed with either *in vitro* translated p70 , p34, or p11. p70 bound to XPA as well as p34, but p11 did not. GST-XPA used in this study was active in an *in vitro* repair assay. Furthermore, the RPA complex was co-immunoprecipitated from HeLa whole cell extracts with XPA by anti-XPA antiserum, suggesting the association of XPA and the RPA complex *in vivo*. To determine the RPA-binding regions of XPA, a series of truncated XPA proteins fused with GST were produced and used for binding to p34, p70, or RPA . In the case of p34, the truncated XPA, which had deletions in C-terminal regions and lacked the 3 N-terminal amino acids, was able to bind the *in vitro* translated p34, while the truncated XPA, which has more than 29 N-terminal amino acids deleted, was unable to bind. However, truncated XPA, which has more than 86 C-terminal amino acids or 137 N-terminal amino acids deleted, was unable to bind the *in vitro* translated p70. These results indicate that amino acids 4-29 and 98-187 of XPA are required for binding to p34 and p70, respectively. When recombinant RPA, which was active in an *in vitro* repair assay, was bound to XPA, it showed a combination of the p34 and p70 patterns. However, deletions of the p34-binding region in XPA (NΔ29, NΔ52, and NΔ97) had little effect on RPA binding, whereas deletion of the p70-binding region (CΔ136 and NΔ137) caused a drastic decrease in RPA binding (Fig 1).

These results indicate that p70 contributed more to the XPA-binding than p34. The binding of p34 to XPA seems to be rather weak and the significance of p34 binding is unknown. However, it may costitute a binding domain in conjunction with p70. It was recently shown that RPA itself preferentially binds to damaged DNA , but complex formation between RPA and UV-damaged DNA was not affected by pre-treatment with dimer photolyase, which removes the cyclobutane pyrimidine dimers from DNA without affecting the content of the (6-4) photoproducts. In fact, RPA bound to substrates with bubble or loop structures. RPA, and probably XPA also, may recognize the single-strandedness resulting from UV lesions. In addition we found that the interaction between XPA and RPA caused a synergistic enhancement of the damaged DNA binding activity of the XPA and RPA

ERCC1 - XPA interaction[8)]

To confirm the interaction of the XPA and ERCC1 proteins, *in vitro* binding assays were also performed. GST-XPA beads bound to the ERCC1 protein in whole cell extracts, and purified XPA protein directly bound to the GST-ERCC1 beads. The effect of ionic strength on the XPA-ERCC1 interaction was then examined. ERCC1 protein binding to GST-XPA beads decreased at higher concentrations of NaCl, but some ERCC1 protein was still bound to XPA beads at 1 M NaCl. These results suggest that some of the interaction between the ERCC1 and XPA proteins is brought by hydrophobic bonds rather than ionic bonds. The interaction between the XPA and ERCC1 proteins might be expected to have biological significance in NER. We therefore examined whether the binding activity of XPA to damaged DNA is affected by the interaction with ERCC1. The UV-damaged DNA binding activity of the XPA protein was markedly increased in the presence GST-ERCC1 protein. The ERCC1 protein alone bound neither UV-damaged nor undamaged double-stranded DNA. When undamaged DNA was used as a probe, little or no enhancement of the DNA binding activity was observed in the concomitant presence of the XPA and ERCC1 proteins. The MF122 protein is a truncated XPA protein which lacks the XPA - ERCC1 interaction regions (and the ability to correct the XPA repair defect) but retains the damaged DNA binding activity. When MF122 was used in place of the wild type XPA protein, ERCC1 did not enhance the UV-damaged DNA binding activity. Group A XP cells expressing a mutant XPA protein with a deletion of seven consecutive glutamic acid residues of the E-cluster region or the C-terminal 48 amino acid residues showed only a marginal recovery of UV-sensitivity, strongly suggesting that the interaction of the XPA protein with the ERCC1 or RPA is indispensable for NER *in vivo*.

Formation of a ternary complex of XPA, RPA, and ERCC1[13)]

As the RPA-binding regions of XPA are different from the ERCC1-binding region (Fig 1), it is possible that XPA binds to both RPA and ERCC1 simultaneously. To verify the formation of a

XPA/RPA/ERCC1 ternary complex *in vitro*, GST-ERCC1 beads were mixed with XPA (N-terminal T7•Tag XPA which was active in an *in vitro* repair assay) and RPA. RPA did not bind to GST-ERCC1 beads in the absence of XPA , indicating that RPA does not bind to ERCC1 directly. But RPA bound to GST-ERCC1 beads in the presence of XPA. These results indicate the formation of a ternary complex *in vitro*. Under these conditions, the interactions between XPA and RPA, or XPA and GST-ERCC1 were also observed.

Binding between XPA and ERCC1 or RPA by the BIAcore system and determination of kinetic parameters[13)]

Direct interactions between ERCC1 or RPA and XPA were monitored using the BIAcore system. GST-ERCC1 and RPA solutions of various concentrations were passed over the XPA-immobilized sensor chip. The association rate constant (k_{ass}) and the dissociation rate constant (k_{diss}) for ERCC1 binding to the surface of XPA were 1.4×10^3 $M^{-1}S^{-1}$ and 3.5×10^{-4} S^{-1}, respectively. The equilibrium dissociation constant (K_D), determined from the ratio of these two kinetic constants (k_{diss}/k_{ass}), was 2.5×10^{-7} M. When RPA solutions (25-400 nM) were examined, the binding was too fast to determine the reliability of the rate constants. The K_D determined by Scatchard analysis of steady state binding levels (Req, relative response at equilibrium) at each concentration was 1.9×10^{-8} M. Binding of RPA and GST-ERCC1 was not observed by the BIAcore system.

Sequential binding of RPA and ERCC1 to XPA[13)]

Next we examined whether RPA and ERCC1 bound in any specific order to XPA. First, 150 nM of RPA was passed over the XPA-sensor chip. The resonance signal increased to 50 RU 300 sec after the injection. Then 150 nM of ERCC1 was injected. The resonance signal increased to 206 RU, and the total resonance signal summed up to 256 RU. These results indicate that the ternary XPA/RPA/ERCC1 complex was formed in the order of injection. Conversely, when ERCC1 was passed over the XPA-sensor chip first, the resonance signal increased to 193 RU 300 sec after the injection. The amount and binding pattern were almost identical to those obtained by ERCC1 injection in the above experiment, indicating that ERCC1 could bind to XPA irrespective of the binding of RPA to XPA. RPA was then injected onto the sensor chip which was pre-bound with ERCC1. Only a little signal increase was observed (7 RU) after the RPA injection, indicating that the pre-binding of ERCC1 to XPA prevents the additional binding of RPA to XPA. When RPA and ERCC1 were injected simultaneously, the resonance signal increased to 268 RU, which was almost equal to that obtained by the sequential bindings of RPA and ERCC1 to XPA. Injection of RPA or ERCC1 alone could not increase this score. All these results suggest that RPA bound to XPA first and then ERCC1 bound to XPA.

We suggest that RPA binds to XPA first and then ERCC1 binds to XPA. ERCC1 forms a stable complex with XPF, and the complex has endonuclease activity and cuts on the 5' side of damaged sites. We have shown that the damaged DNA binding activity of XPA is increased by binding to

ERCC. Therefore we believe that ERCC1 assembled with XPA/RPA at the damaged sites and ERCC1/XPA interactions recruit the ERCC1/XPF endonuclease to damaged sites in the DNA. On the other hand, XPG is reported to bind to RPA, suggesting that XPA/RPA also attracts the XPG nuclease to the 3' side of damaged DNA. All these results suggest that pre-assembled XPA/RPA congregates at damaged sites and becomes the core of the incision protein complex which efficiently advances the NER reaction. It has been reported that in *Saccharomyces cerevisiae* Rad14 binds more tightly to the Rad1/Rad10 complex than to Rad1 and Rad10 proteins alone, and a higher level of incision of UV-damaged DNA is achieved with the pre-assembled Rad1/Rad10/Rad14 complex. These results demonstrate that quantitative measurements of interactions between NER proteins can reveal sequential protein associations during repair, and aid in elucidating the mechanism of NER as the repair complex assembles at sites of damage in DNA.

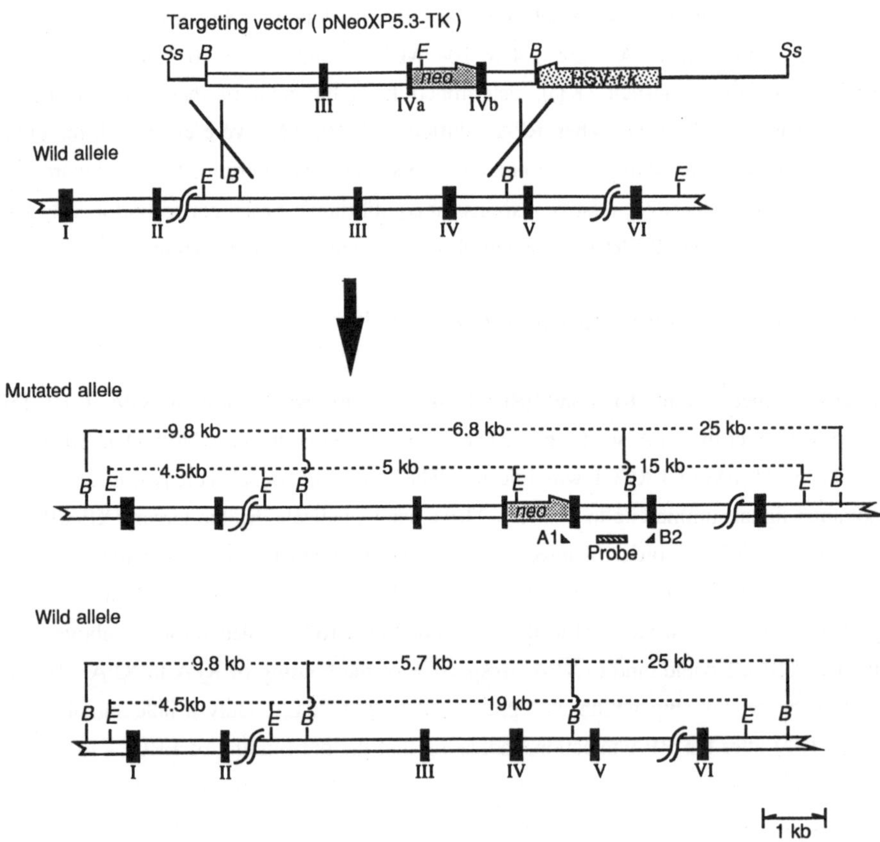

Fig 2. Targeted inactivation of the *XPA* gene in ES cells. Thin line represents the vector backbone; closed boxes represent exons of the *XPA* gene; and open boxes represent the introns and flanking regions. The restriction enzyme sites shown are B, *Bam*HI; E, *Eco*RI; and Ss, *Ssp*I.

XPA - deficient mice[9, 12)]

To elucidate the *in vivo* function and molecular basis of group A XP phenotype, we established *XPA*-deficient mice. Mice with a disrupted *XPA* allele were generated by insertion of neomycin (*neo*) cassette sequences into exon 4 of the *XPA* gene using embryonic stem (ES) cell techniques (Fig 2). The heterozygous (+/-) mice did not display any abnormal phenotype compared to their wild type (+/+) littermates as judged by physical observation. The (+/-) mice were intercrossed to obtain the homozygous (-/-) mice. The frequency of the weaned pups with (+/+), (+/-) and (-/-) *XPA* genotypes was 24.5%(179 mice), 49.8%(363 mice) and 25.7% (187 mice), respectively, indicating that the disruption of the *XPA* gene by our method does not result in embryonic lethality. The (-/-) mice showed neither obvious physical abnormalities nor pathological alterations and were fertile. To confirm that no *XPA* gene products were being synthesized in the (-/-) mice, western blot analysis of the splenocyte extracts using anti-mouse XPA protein antibody was performed. Approximately half of the XPA protein detected in the (+/+) type mice was produced in the (+/-) mice, whereas no XPA proteins were detected in the (-/-) mice. To confirm that the targeting of the *XPA* gene disrupted NER capacity of mice, the colony forming ability of fibroblasts derived from (-/-), (+/-) and (+/+) newborn mice was compared after various doses of ultraviolet light (UV) (254 nm)-irradiation. The (-/-) fibroblasts were hypersensitive to UV-irradiation, while the (+/-) and (+/+) fibroblasts were resistant to the UV-irradiation. The UV survival curve of fibroblasts from (-/-) mice was almost the same as that from group A XP patients with typical clinical phenotype [D_{37} = 0.3 J/m^2 for the (-/-) mouse fibroblasts; 0.2 - 0.4 J/m^2 for the cells from group A XP patients]. Repair synthesis of DNA during NER (so-called unscheduled DNA synthesis : UDS) in the (-/-), (+/-) and (+/+) fibroblasts was 3± 2, 61±14, and 58± 12 (mean grain number ± standard deviation), respectively. Thus, the UV-induced UDS was absent in the (-/-) fibroblasts. We also compared the kinetics of removal of major UV induced DNA lesions in these fibroblasts by ELISA using monoclonal antibodies against cyclobutane-type thymine dimers and (6-4) photoproducts. Neither thymine dimers nor (6-4) photoproducts were removed in the (-/-) fibroblasts but both were removed in the (+/+) and (+/-) fibroblasts. Thus, we concluded that the (-/-) mouse was defective in NER as typical human group A XP patients.

We then examined UVB-induced skin carcinogenesis in these mice. The shaved back skins of 15 (-/-), 22 (+/-) and 15 (+/+) mice were irradiated three times a week with UVB at a dose of 2000 J/m^2 per exposure. Areas other than shaved back skin were shielded from UVB. After 1 2 weeks of UVB irradiation, all the (-/-) mice developed skin ulcers in the exposed areas, also indicating that the (-/-) mice are highly susceptible to UVB. The ulcers were gradually surrounded by poorly defined areas of hyperkeratotic adherent scale. After 20 weeks of UVB-irradiation, the first skin tumor was detected in the (-/-) mice as a nodule in the area of the hyperkeratotic scale, and by 34 weeks of UVB-irradiation every (-/-) mouse developed such tumors. On the other hand, the (+/+) and (+/-) mice did not develop any significant skin abnormalities during the UVB-irradiation experiments. All the (-/-), (+/-) and (+/+) mice were killed for histological analysis at week 39. All of the skin tumors in the (-/-) mice were histologically diagnosed as squamous cell carcinoma (SCC). On the other hand, the UVB-irradiated skins of (+/-) and (+/+) mice were nearly normal.

These findings indicate that the *XPA* (-/-) mice are highly susceptible to UVB-induced skin carcinogenesis, that the XPA protein protects mice from carcinogenesis initiated by UV irradiation, and that the *XPA*-deficient mice may provide a useful model to study the high incidence of skin carcinogenesis in group A XP. On the other hand, many of the group A XP patients develop a number of neurological complications such as microcephaly, progressive mental deterioration and sensorineural deafness. The major pathological findings in the nervous system of group A XP patients are loss of neurons and neurodegeneration, particularly in the cerebral cortex and cerebellum. However, pathological analysis of neural tissues from the *XPA*-deficient mice revealed no neurodegeneration or gliosis. The reason for this difference remains unknown at present.

References

1) Tanaka, K., Satokata, I., Ogita, Z., Uchida, T., and Okada, Y., *Proc. Natl. Acad. Sci. USA*, **86**, 5512-5516, 1989.

2) Tanaka, K., Miura, N., Satokata, I., Miyamoto, I., Yoshida, M. C., Satoh, Y., Kondo, S., Yasui, A., Okayama, H. and Okada, Y., *Nature*, **348**, 73-76, 1990.

3) Miura, N., Miyamoto, I., Asahina, H., Satokata, I., Tanaka., K. and Okada, Y., *J. Biol. Chem.*, **266**, 19786-19789, 1991.

4) Shimamoto, T., Kohno, K., Tanaka, K. and Okada, Y., *Biochem. Biophys. Res. Comm.,* **181**, 1231-1237, 1991.

5) Miyamoto, I., Miura, N., Niwa, H., Miyazaki, J. and Tanaka, K., *J. Biol. Chem.*, **267**, 12182-12187, 1992.

6) Asahina, H., Kuraoka, I., Shirakawa, M., Eugene H. Morita, Miura, N., Miyamoto, I., Ohtsuka, E., Okada, Y. and Tanaka, K., *Mutat. Res.*, **315**, 229-237, 1994.

7) Matsuda, T., Saijo, M., Kuraoka, I., Kobayashi, T., Nakatsu, Y., Nagai, A., Enjoji, T., Masutani, C., Sugasawa, K., Hanaoka, F. Yasui, A. and Tanaka, K., *J. Biol. Chem.*, **270**, 4152-4157, 1995.

8) Nagai, A., Saijo, M., Kuraoka, I., Matsuda, T., Kodo, N., Nakatsu, Y., Mimaki, T., Mino, M., Biggerstaff, M., Wood, R. D., Seijbers, A., Hoeijmakers, J. H. J. and Tanaka, K., *Biochem. Biophys. Res. Comm.*, **211**, 960-966, 1995.

9) Nakane, H., Takeuchi, H., Yuba, S., Saijo, M., Nakatsu, Y., Murai, H., Nakatsuru, Y., Ishikawa, Hirota, S., Kitamura, Y., Kato, Y., Tsunoda, Y., Miyauchi, H., Horio, T., Tokunaga, T., Matsunaga, T., Nikaido, O., Nishimune, Y., Okada, Y., and Tanaka, K., *Nature*, **377**, 165-168, 1995.

10) Kuraoka, I., Morita, E. H., Saijo, M., Matsuda, T., Morikawa, K., Shirakawa, M. and Tanaka, K., *Mutat. Res.*, **362**, 87-95, 1996.

11) Morita, E. H., Ohkubo, T., Kuraoka, I., Shirakawa, M., Tanaka, K., Morikawa, K., *Genes to Cells*, **1**, 437-442, 1996.

12) Miyauchi, H., Tanaka, K. and Horio, T., *J. Invest. Dermat.*, **107**, 343-348, 1996.

13) Saijo, M., Kuraoka, I., Masutani, C., Hanaoka, F. and Tanaka, K., *Nucl. Acid Res.*, in press .

18. Characterization of *xpg*-deficient Mice

Yoshi-nobu HARADA,[1] Naoko SHIOMI,[1] Masahito IKAWA,[2] Masaru OKABE,[2] Seiichi HIROTA,[3] Yukihiko KITAMURA[3], Manabu KOIKE[1] and Tadahiro SHIOMI[1]

[1]The Genome Research Group, National Institute of Radiological Sciences, Chiba 263, Japan; [2]Research Institute for Microbial Diseases, Osaka University, Osaka 565, Japan; [3]Department of Pathology, Osaka University Medical School, Osaka 565, Japan

Xeroderma pigmentosum (XP) and Cockayne syndrome (CS) are two rare inheritable disorders, which are characterized by a clinical and cellular hypersensitivity to the ultraviolet (UV) component of the sunlight spectrum.[1,2)] Cell fusion studies have revealed eight complementation groups in XP (A-G, and an XP-variant form) and two complementation groups in CS (A and B).[3,4)] XPs and CSs are demonstrated to be caused by defects in an early step(s) of nucleotide excision repair (NER) pathway. At present, genes that complement six of eight XP groups (A-E and G) and two of CS groups (CS-A and CS-B) have been molecularly cloned and characterized. To investigate further unknown function of the genes involved in NER, knockout mice have been generated using embryonic stem cell technology for *ERCC-1* ,[5)] *XPA*[6,7)] and *XPC*.[8)] Increased sensitivity to chemical carcinogens and UV irradiation has been observed in these knockout mice. The *ERCC-1*-deficient mice were severely runted, died before weaning, and had elevated levels of p53 and liver nuclear abnormalities. Some characteristic defects often found in XP patients such as neurological abnormalities have not been observed in any of these knockout mice.

XP-G, the seventh complementation group, is one of the most rare and phenotypically heterogeneous groups. XP-G patients vary from having very mild cutaneous abnormalities to severe dermatological impairments, and in a few cases a combination of clinical hallmarks of XP-G and CS has been reported.[9)] The XPG gene encodes an acidic protein with predicted molecular mass of 133 kDa that shares two regions of extensive homology with the yeast DNA repair protein *RAD2*.[10-12] The XPG protein is reported to function as a magnesium-dependent single-stranded DNA endonuclease.[3,13]

To generate a mutant allele of *xpg*, mouse genomic clones containing several exons of the *xpg* gene were isolated from a lambda DASH II (Stratagene) phage library constructed with DNA from the D3 (ES cell line derived from mouse strain 129/Sv). A pMER5/TV2 targeting vector was designed to generate an insertional mutation in one exon of the mouse *xpg* gene. Targeted clones with the predicted insertional mutation were identified with a probe 3' to the homology region and PCR analyses. Targeted ES cells were injected into C57BL/6 blastocysts to generate chimeric mice which transmitted the mutant allele (designated as xpg^{mt2}) to F1 offspring as identified by PCR and Southern blot analyses. To examine the effects of the insertional mutation on expression of the *xpg* gene, total RNA from newborn mice was analyzed by Northern blotting. No stable (intact or truncated) *xpg* transcript was detected in the -/- (abbreviation for xpg^{mt2}/xpg^{mt2}) homozygous mice using the *xpg* cDNA as a probe. In heterozygous (+/-) mice, the *xpg* mRNA content was approximately half that in wild-type (+/+) mice. These findings indicate that disruption of the *xpg* gene was effective.

The heterozygous mice were interbred in an attempt to obtain mutant homozygous mice. Of 163 pups born, 35 (21.5%) mice were severely runted and died between 0 and 23 days *post partum*. DNAs were successfully collected from 31 of 35 dead mice. Southern blot and PCR analyses of these DNAs revealed that all 31 mice were homozygous for the targeted *xpg* gene. The 128 survivors were either wild-type (39, 23.9%) or heterozygotes (89, 54.6%). The three genotypes (of embryos at 16 and 21 days *post coitum)* were obtained at the expected Mendelian ratio (9 wild type, 19 heterozygotes, 10 mutant homozygotes) and no sexual bias was observed, indicating that disruption of the *xpg* gene by our method does not affect embryonic viability before birth. Although the size of the mutant homozygotes at birth seemed to be almost the same as those of wild-type and heterozygote pups, growth of the *xpg* mutant homozygotes was severely inhibited thereafter.The average body weight of normal mice (wild-type and heterozygote) and mutant homozygote mice at 16 days was 8.3g and 3.8g, respectively. Most mutant homozygotes died between 12 and 19 days *post partum*. All mutant homozygotes died before weaning. The adult heterozygous mice were indistinguishable from their wild-type littermates in both appearance and fertility. The wild-type and heterozygous mice showed neither obvious physical abnormalities nor pathological alterations.

Primary fibroblasts were isolated from newborn mice of all three *xpg* genotypes and their susceptibilities to UV irradiation were examined by measuring colony forming ability after exposure to various doses of UV. The fibroblasts from *xpg* mutant homozygotes proliferated in culture as rapidly as wild-type cells, and plating efficiencies were almost the same among these fibroblasts. The fibroblasts from -/- mice were hypersensitive to UV (254 nm) irradiation. Cells from heterozygotes were as resistant to UV irradiation as wild-type cells. The UV survival curve for -/- fibroblasts was almost the same as that for cells from a severe XP-G patient.[10)]

Next we investigated the cause of severe growth inhibition and premature death in the mutant homozygotes. The homozygous mutant mice seemed to cling similarly to the teats of their dams as the normal mice, although the growth of the mutants was severely retarded. After 10 days, the mutant homozygotes became weak and emaciated, and their activity was relatively low. Histological analysis has been carried out on several organs of mice. Cells of the various organs in -/- mice were smaller than those in +/+ or +/- mice. For example, liver cells in mutant mice were markedly smaller than those in heterozygous mice. The spleen was very small in all -/- mice. The stomachs of wild-type and heterozygous mice contained milk and food, whereas those of -/- mice contained little. The intestines of -/- mice were a very clear yellow color, indicating the presence of little milk. There were many gas bubbles in the intestines of -/- mice. These observations suggest that the mutant homozygotes did not (or could not) take sufficient food and were in a state of starvation atrophy.

It is still unknown why the *xpg* mutant mice were in a state of starvation atrophy. The observations in this study suggest that the *xpg*-deficient mice did not (or could not) take sufficient food, were in a state of starvation atrophy and emaciated, and finally died before weaning. These symptoms must be caused by the defects of the *xpg* gene products. The *xpg* gene may have another function which is not involved in the NER pathway.

It is possible that the *xpg*-deficient mice also had CS, since it has been reported that some severely affected XP-G patients also exhibit symptoms of CS. The *xpg*-deficient mice generated in this study could be a model for severely affected XP-G patients, and the severe runting and premature death of the *xpg*-deficient mice could be the equivalent of growth inhibition and relatively short life span of CS patients.

In this study, the *xpg* gene was demonstrated to be necessary for normal growth in mice. The *xpg* mutant mice are a very useful model for studying the biological functions of the murine *xpg* and human *XPG* genes.

References

1. Cleaver, J.E., and K.H. Kraemer. In *The Metabolic Basis of Inherited Disease* (eds. C.R. Scriver, A.L. Beaudet, W.S. Sly, and D. Valle), McGraw-Hill, New York, 1989, pp. 2949-2971.
2. Friedberg, E.C., G.C. Walker., and W. Siede. In: *DNA Repair and Mutagenesis,* American Society for Microbiology, Washington, D.C., 1995, pp. 655-658.
3. Aboussekhra, A., and R.D. Wood. *Curr. Opin. Genet. Dev.*, 1994, **4**: 212-220.
4. Venema, J., L.H. Mullenders., A.T. Natarajan., A.A. van Zeeland, and L.V. Mayne. *Proc. Natl. Acad. Sci. USA* , 1990, **87**: 4707-4711.
5. McWhir, J., J. Selfridge., D.J. Harrison., S. Squires., and D.W. Melton. *Nat. Genet.,* 1993, **5**: 217-224.
6. Nakane, H., S. Takeuchi., S. Yuba., M. Saijo., Y. Nakatsu., H. Murai., Y. Nakatsuru., T. Ishikawa., S. Hirota., Y. Kitamura., Y. Kato., Y. Tsunoda., H. Miyauchi., T. Horio., T. Tokunaga., T. Matsunaga., O. Nikaido., Y. Nishimune., Y. Okada., and K. Tanaka. *Nature* , 1995, **77**: 165-168.
7. de Vries, A., C.T. van Oostrom., F.M. Hofhuis., P.M. Dortant., R.J. Berg., F.R. de Gruigl., P.W. Wester., C.F. van Kreijl., P.J. Capel., H. van Steeg., and S.J. Verbeek. *Nature* , 1995, **377**: 169-173.
8. Sands, A.T., A. Abuin., A. Sanchez., C.J. Conti., and A. Bradley. *Nature,* 1995, **377**: 162-165.
9. Vermeulen, W., J. Jaeken., N.G. Jaspers., D. Bootsma., and J.H.J. Hoeijmakers. *Am. J. Hum. Genet.* 1993, **53**: 185-192.
10. Shiomi, T., Y.-N. Harada., T. Saito., N. Shiomi., Y. Okuno., and M. Yamaizumi. *Mutat. Res.,* 1994, **314**: 167-175.
11. Scherly, D., T. Nouspikel., J. Corlet., C. Ucla., A. Bairoch., and S.G. Clarkson. *Nature* , 1993, **363**: 182-185.
12. Harada, Y.-N., Y. Matsuda., N. Shiomi., and T. Shiomi. *Genomics* , 1995, **28**:

59-65.

13. O'Donovan, A., D. Scherly., S.G. Clarkson., and R.D. Wood. *J. Biol. Chem.*, 1994, **269**: 15965-15968.

19. Transcription-coupled Repair as a Biodefence Mechanism

Gijsbertus T.J. van der HORST, Harry van STEEG[1], Rob J.W. BERG[2], Kiyoji TANAKA[3], Errol FRIEDBERG[4], Dirk BOOTSMA and Jan H.J. HOEIJMAKERS

Dept. of Cell Biology & Genetics, MGC, Erasmus University, PO Box 1738, 3000 DR Rotterdam, [1)] Dept. LCM, RIVM, Bilthoven, [2)] Dept. of Dermatology, University of Utrecht, Utrecht, The Netherlands, [3)] Inst. Molecular & Cellular Genetics, Osaka University, Osaka, Japan, [4)] Dept. of Pathology, University of Texas Southwestern Medical Centre, Dallas, USA

Nucleotide Excision Repair Pathways

Maintenance of genetic information is of vital importance to all living organisms. Accumulation of damage in the DNA can lead to mutations that cause inborn defects, cell death or cellular malfunctioning, including uncontrolled growth. To counteract the deleterious effects of environmental and endogenous genotoxic agents, cells are equipped with a complex network of repair systems with complementary substrate specificity. One of the systems that keep the DNA under continuous surveillance is nucleotide excision repair (NER). It is responsible for the elimination of a wide class of structurally unrelated lesions such as UV-induced cyclobutane pyrimidine dimers (CPD) and (6-4) photoproducts (6-4PP), chemically induced bulky adducts and intra-strand crosslinks (for review see Hoeijmakers, 1994; Friedberg et al., 1995).

The analysis of cloned NER genes and development of *in vitro* repair assays and reconstitution systems with purified proteins have contributed to the unraveling of the complex multi-step NER process (for reviews see Hoeijmakers, 1994 and Wood, 1996). The reaction requires the concerted sequential action of some 30 proteins and involves the following steps: recognition of the DNA lesion, local opening of the double helix around the injury, incision of the damaged strand on both sides of the lesion, release of a 27-29mer damage-containing oligonucleotide, gap-filling DNA synthesis and ligation. From the work of the group of Hanawalt it became evident that NER consists of two subpathways (reviewed by Hanawalt, 1994). The Global Genome Repair (GGR) subpathway is responsible for the removal of lesions from the entire genome. However, a major drawback of the system is that certain types of damage (like UV-induced CPDs) are less well recognized and accordingly less efficiently repaired. To avoid that such lesions interfere with transcription by stalling

RNA polymerase II, cells have developed the Transcription-Coupled Repair (TCR) subpathway. This system directs the repair machinery preferentially to the template strand of actively transcribed DNA and operates as a fast backup system for lesions that are slowly repaired by GGR.

Human NER syndromes and Mouse models

The phenotypical consequence of defective NER in mammals is illustrated by a group of genetically heterogeneous photo-sensitive disorders (Table 1). *Xeroderma pigmentosum* (XP; complementation groups A to G) is characterized by cutaneous abnormalities and increased skin cancer predisposition, often in combination with progressive neurological degeneration (Bootsma et al, 1996). Most XP patients have defective GGR and TCR; in XP-C only GGR is affected. In *Cockayne syndrome* (CS; complementation groups A and B) only TCR is impaired. Patients present with postnatal growth failure, impaired sexual development, and severe neurological dysfunction resulting from dysmyelination (Nance and Berry, 1992). Importantly, CS is not known to be associated with skin cancer predisposition. A number of XP patients (XP-B, D and G) display clinical symptoms of CS and XP. Patients with *Trichothiodystrophy* (TTD; 3 complementation groups TTD-A, XP-B and XP-D) are defective in GGR and TCR. The hallmark of TTD is brittle hair and nails in addition to ichthyosis and the symptoms of CS (Itin and Pittelkow, 1990).

The marked differences in clinical effects of defective NER in the three photosensitive disorders are difficult to explain. First, why is the NER defect in CS and TTD not associated with skin cancer predisposition? The assumption that GGR rather than TCR is involved in the suppression of mutagenic and carcinogenic events may rationalise the absence of cancer in CS but does not explain why TTD patients are not cancer-prone. Alternatively it has been proposed that transcription-blocking lesions, resulting from defective TCR, trigger apoptosis to eliminate potentially mutagenic cells

Table 1 Properties of XP, CS and TTD complementation groups

compl. group	affected NER GGR	TCR	skin cancer	neurologic abnorm.	disease
XP-A	-	-	+	+ +	XP
XP-B	-	-	+/-	+ +/+	XP/CS,TTD
XP-C	-	+	+	-	XP
XP-D	-	-	+/-	+ +/±	XP,XP/CS,TTD
XP-E	?	?	+/-	-	XP
XP-F	-	-	+/-	-/±	XP
XP-G	-	-	+/-	+ +/+	XP,XP/CS
TTD-A	-	-	-	+	TTD
CS-A	+	-	-	+ +	CS
CS-B	+	-	-	+ +	CS

(Ljungman and Zhang, 1996). Although this scenario may hold for CS and TTD, it does not explain why XP patients with a combined defect in GGR and TCR are skin cancer susceptible. It is clear that the precise role of NER proteins in induction or suppression of carcinogenesis is far from understood. Furthermore, the CS and TTD features are difficult to explain in terms of a NER defect. Unexpectedly, the XPB and XPD proteins were shown to be subunits of TFIIH, a basal transcription factor involved in repair. The dual function of these proteins in transcription and repair opens the possibility that some of the CS and TTD symptoms are due to a subtle defect in transcription of a specific subset of genes and has lead to the transcription syndrome concept (Bootsma and Hoeijmakers, 1993). On the basis of the CS involvement it is inferred that mutations in XPG, CSA and CSB may also affect basal transcription.

The generation of animal models for photosensitive disorders will contribute to the understanding of the function of (defective) NER proteins in mutagenesis, carcinogenesis and the onset of repair and/or transcription symptoms. Recently, mouse models for XPA and XPC have been shown to reliably reflect the skin cancer predisposition of the corresponding human disorder: repair- deficient mice readily develop skin tumors when exposed to UV- light or chemical carcinogens (de Vries et al., 1995; Nakane et al., 1995; Sands et al., 1995). However, XPA-deficient mice failed to show the neurological symptoms characteristic for human XP-A. Animal models for Cockayne syndrome and trichothiodystrophy have not been reported yet.

Table 2 UV Repair characteristics of human and rodent CS cells

	Human	Rodent
UV-sensitivity (at D_{10})	3-3.5x↑	3-3.5x↑
RNA synthesis recovery after UV	defect	defect
Repair synthesis (UDS)	normal	normal
Repair of 6-4PP	normal	normal[a]
Repair of CPDs (% of wt, 16 hr post-UV)	≈60%[b]	<10%[b]

[a] determined in CHO line UV61

[b] repair by GGR only (in rodents much lower efficiency than in man)

Mouse models for Cockayne syndrome

For both CS complementation groups, the corresponding genes have been isolated by transfection-correction cloning. The CSB gene encodes a 1493 aa protein with highly conserved SWI/SNF2-like DNA-dependent ATPase domain (Troelstra et al., 1992). The CSA gene product is a 396 aa protein with conserved WD-repeats and has been claimed to interact with the CSB protein (Henning et al., 1995). On the basis of the function of other members of the SWI/SNF protein family it is envisaged that the CSB protein (or the CSA/CSB protein complex) may be involved in chromatin remodeling and/or displacement of the stalled RNA- polymerase to make the lesion accessible to the repair machinery.

Animal models for CS-A and CS-B have been obtained by targeted disruption of the corresponding mouse genes. The CSA gene was interrupted by insertion of a selection marker and mice heterozygous for the targeted allele have been obtained (unpublished results). We are currently awaiting the birth of homozygous mutant mice for further analysis. An animal model for CS-B was generated by mimicking a mutated allele of a known CS-B patient in the mouse CSB gene (van der Horst et al., 1997). Homozygous mutant mice, although synthesizing low levels of the CSB-Neo fusion transcript, do not produce the wildtype CSB protein and accordingly can be considered as functional null- mutants. Repair parameters have been analyzed in mouse embryonic fibroblasts (Table 2). Analogous to fibroblasts from patients with CS-B, CSB-deficient mouse cells display UV-sensitivity, disturbed RNA synthesis recovery, normal GGR (measured by UDS), and defective preferential repair of CPDs in the transcribed strand of an active gene. Moreover, CSB-deficient mice have been proven photo-sensitive: exposure of the shaven skin to UV-light results in the development of an erythemic reaction and epidermal hyperplasia. Our data demonstrate that CSB-deficient mice are a bona fide model for the TCR defect in CS.

Cockayne syndrome and cancer predisposition

Skin cancer has never been described for CS (Nance and Berry, 1992). To verify the absence of cancer predisposition in CS, we have subjected CSB-deficient mice to carcinogenesis protocols as performed for XPA and XPC-deficient mice (de Vries et al., 1995; Nakane et al., 1995; Sands et al., 1995). Unexpectedly, and in striking contrast to human CS, CSB-deficient mice appear clearly prone to skin cancer when exposed to physiologically relevant doses of ultraviolet light. A tentative comparison of carcinogenic events in XPA- and CSB-deficient mice reveals a longer latency time and a higher cumulative UV-dose for squamous cell carcinoma development in the latter. This suggests

Table 3 Comparison of human and rodent Cockayne syndrome

Clinical feature	Human CS	Rodent CS
Repair related:		
Photosensitive skin	+	+
Photophobia	+	+
Skin cancer predisposition		
UV-induced	not reported	+
Chemically-induced	not reported	+
Other:		
Life span	reduced	normal
Growth retardation	severe	mild[1]
Neurological abnormalities	severe	mild[2]
Reproduction	infertility	normal

[1]) mainly observed in ♂ mice

[2]) except for severe retinopathy

that in the rodent situation the cancer predisposition in CS may be less pronounced than in XP. Currently we are backcrossing XPA- and CSB-deficient mice in the same (hairless) background to allow a proper comparison of skin cancer susceptibility. Similarly, we have shown that CSB-deficient mice develop skin tumors when exposed to the chemical carcinogen DMBA. In conclusion, our data clearly demonstrate that in rodents GGR alone is not sufficient to prevent skin cancer induction and that TCR also acts as a biodefence mechanism against mutagenic and carcinogenic events. In addition, it appears that the proposed protective effect of transcription-blocking lesion induced apoptosis (Ljungman and Zhang, 1996), if present, is not sufficient to prevent carcinogenic events.

The observed skin tumor susceptibility in rodent CS may have implications for human CS. It is possible to explain the absence of skin cancer in human CS on the basis of (species- specific) differences in the efficiency of GGR: whereas in absence of TCR rodent cells hardly repair CPDs in the transcribed strand of active genes, human cells can still repair up to 60% of the lesions by GGR (van Hoffen et al., 1993). Apparently, in man GGR on its own is sufficiently efficient to prevent or at least slow down mutagenic events whereas in rodents it is not. It is tempting to associate the need for a more developed GGR subpathway in man with the absence of a furred skin and the high activity during daylight. Alternatively it is conceivable that CS patients have an hitherto unnoticed cancer predisposition, although less pronounced than in XP. Given the notion that the average age of death in human CS is 12 years and extrapolating from the fact that at least in rodent models tumor development in CS may proceed slower than in XP, CS patients may not live long enough to actually get skin cancer.

Origin of clinical features in Cockayne syndrome

In addition to photosensitivity, typical clinical CS features involve poor growth, impaired sexual development and neurologic abnormalities, including disturbed motor function, sensorineural hearing loss and ocular abnormalities. However, CSB-deficient mice neither show a reduced life span nor fertility problems; only careful examination of mutant mice uncovered mild CS-like features like a small reduction in body weight in male mice, slightly disturbed motor coordination and indications of hearing loss. Histological examination of tissues revealed no pathological changes, except for the eyes: a consistent severe retinopathy (consisting of loss of photoreceptors) was found in all older (>1 year) CSB-deficient mice examined. The absence of retinal abnormalities in totally NER-defective XPA-deficient mice as well as GGR defective XPC deficient mice (de Vries et al., 1995; Nakane et al., 1995; Sands et al., 1995) suggests an additional function for the CSB protein outside the context of NER.

A remarkable observation was made when XPC/CSB double-deficient mice (defective TCR and GGR) were generated for comparison with XPA-deficient mice (also totally NER- deficient). To our surprise XPC/CSB-deficient mice show a severe growth retardation of 75% and die around day 18. Similarly, XPA/CSB double-deficient mice are severely runted and die before weaning. In contrast, apart from the NER defect, XPA/XPC double deficient mice appear normal. These observations suggest that the phenotypical effect of the CSB-deficiency becomes exaggerated in the absence of

GGR and point to accumulation of endogenous damage as a contributing factor to CS symptoms. Provided that accumulation of critical levels of endogenous damage in a specific set of genes is indeed a contributing factor to the onset of CS symptoms, the expected absolute rather than life span related time scale of damage accumulation may explain the mild CS phenotype of CSB-deficient mice. In patients with severe CS (such as patients with combined XP/CS) the neurodevelopmental symptoms appear within the first two years of life, whereas classical CS (as in the case of most CS-B patients) is even diagnosed later (Nance and Berry, 1992). Thus, the time required for damage accumulation in human CS approaches or even exceeds the usual life span of the mouse.

Concluding remarks

The development of transgenic and knockout mouse technology has enabled the generation of mouse models for human disorders. Those models are important tools for studying the etiology and pathophysiology of human disorders as animals can easily be subjected to experimental protocols and detailed analyses not applicable to humans. We have generated mouse models for both complementation groups of Cockayne syndrome and unexpectedly found a cancer predisposition in the rodent equivalent of CS-B. Thus, in addition to defective GGR (as in XP) defective TCR can also result in tumorigenic events. The CSB-deficient mouse model allows us to study the underlying mechanism in more detail. In addition to properly compare the cancer susceptibility in XP and CS mice, we are currently focusing our attention on mutagenesis, apoptosis, immunological parameters and other aspects that determine carcinogenesis. Moreover, one of the advantages of yeast genetics - the generation of double mutants - has now become available for mammalian systems. By generating double mutant NER mice we have shown that accumulation of endogenous damage may somehow contribute to the onset of clinical features in CS. At this moment one of the major challenges will be to find the responsible type of damage and to elucidate the role of the CSB protein in transcription / TFIIH functioning.

Acknowledgements

We are grateful Dr. Alain van Gool and Mr. Jan de Wit for their help in the analysis of mutant mice. We also thank our colleagues of the DNA repair group of the Medical Genetics Centre for stimulating discussions. Research in our group is supported by the Dutch Scientific Organization (NWO) and grants from the Dutch Cancer Society (KWF) and the Human Frontier Science Program (HFSP).

References

Bootsma, D., Cleaver, J.E., Kraemer, K.H., and Hoeijmakers, J.H.J. (1996). In The Metabolic Basis of Inherited Disease, Vol. 3, C.R. Scriver, A.L. Beaudet, W.S. Sly, and D.Valle, eds. (New York: McGraw-Hill), in press

Bootsma, D., and Hoeijmakers, J.H.J. (1993). Nature, 363, 114- 115.

de Vries, A., van Oostrom, C.T.M., Hofhuis, F.M.A., Dortant, P.M., Berg, R.J.W., de Gruijl, F.R., Wester, P.W., van Kreijl, C.F., Capel, P.J.A., van Steeg, H., and Verbeek, S.J. (1995). Nature, 377, 169-173.

Friedberg, E.C., Walker, G.C., and Siede, W. (1995). DNA repair and mutagenesis. ASM Press Washington DC.

Hanawalt, P.C. (1994). Science, 266, 1957-1958.

Henning, K.A., Li, L., Iyer, N., McDaniel, L.D., Reagan, M.S., Legerski, R., Schultz, R.A., Stefanini, M., Lehmann, A.R., Mayne, L.V. and Friedberg, R.C. (1995). Cell, 82, 555-564.

Hoeijmakers, J.H.J. (1994) Eur. J. Cancer, 30A, 1912-1921.

Hoeijmakers, J.H.J., Egly, J.M., and Vermeulen, W. (1996). Curr. Opin. Genet. Dev. 6, 26-33.

Itin, P.H., and Pittelkow, M.R. (1990). J. Am. Acad. Dermatol., 22, 705-717.

Ljungman, M., and Zhang, F. (1996). Oncogene 13, 823-831.

Nance, M.A., and Berry, S.A. (1992). Am. J. Med. Genet., 42, 68- 84.

Nakane, H., Takeuchi, S., Yuba, S., Saijo, M., Nakatsu, Y., Ishikawa, T., Hirota, S., Kitamura, Y., Kato, Y., Tsunoda, Y., Miyauchi H., Horio, T., Tokunaga, T., Matsunaga T., Nikaido, O., Nishimune, Y., Okada, Y., and Tanaka, K. (1995). Nature, 377, 165-168.

Sands, A.T., Abuin, A., Sanchez, A., Conti, C.J., and Bradley, A. (1995). Nature, 377, 162-165.

Troelstra, C., van Gool, A., de Wit, J., Vermeulen, W., Bootsma, D., and Hoeijmakers, J.H.J. (1992). Cell, 71, 939-953.

van der Horst, G.T.J., van Steeg, H., Berg, R.J.W., de Wit, J., van Gool, A.J., Weeda, G., Morreau, H., Beems, R.H., van Kreijl, C.F., de Gruijl, F.R., Bootsma, D., and Hoeijmakers, J.H.J. (1997). submitted.

van Hoffen, A., Natarajan, A.T., Mayne, L.V., van Zeeland, A.A., Mullenders, L.H.F., and Venema, J. (1993). Nucleic Acids Res., 21, 5890-5895.

Wood, R.D. (1996). Ann. Rev. Biochem. 65, 135-167.

Index